CW00501749

THE GROWN-UPS BOOK OF RISK

WHY ##IT HAPPENS

I hope that you enjoy the book. With Best Wishes, Omar Malik.

Dec 08

OMAR MALIK

This book is dedicated to all decent and honest people everywhere. May the souls of the others, who are not decent and honest, rest in peace, and soon.

Published in Great Britain by New Insight Press, PO Box 3500,
Malmesbury SN19 1AF
www.newinsightpress.co.uk

A CIP catalogue record for this title is available from the British Library

ISBN 978-0-9560224-0-0

Designed and typeset by Bookcraft Ltd, Stroud, Gloucestershire

Printed and bound in Great Britain by CPI Antony Rowe, Chippenham, Wiltshire

CONTENTS

CHAPTER 2 MORE ABOUT WHY THINGS GO WRONG

CHAPTER 3 A TOE IN THE WATERS OF RISK MANAGEMENT

PART III: THE PREVENTION OF FAILURE

CHAPTER 8 FAILURE PATH® ANALYSIS AND ACCIDENT PREVENTION

PREFACE

Life is not one long orgasm. Pity, but there it is.

So what is it? It is a search for self-fulfilment, and that is where the simple answers end. In unhappy countries self-fulfilment is being alive and well at the end of the day. In more civilised lands, safety and nourishment are in little doubt. To us, the fortunate, self-fulfilment is the task of trying to extract what we want from our allotted time on earth. Our success in achieving this is measured by our happiness. Unfortunately the world is so arranged that a small amount of unhappiness can easily overwhelm a large helping of happiness. Everyone has their individual view of what constitutes self-fulfilment: some worthy, some less so. However, all will probably agree that the essential first step in the pursuit of happiness is the avoidance of unhappiness.

This book is about the management of the risk of unhappiness. It is a risk which faces us all. It is a risk which arises from our well-being and our health; they may be threatened. It arises from our friends and our families; they may suffer harm. It arises from our activities; they may go wrong. It arises from our assets; they may be damaged or lost. Unhappiness is a product of adverse events. Life is about managing the risks of unhappiness; it is about the avoidance of adverse events.

The management of risk is now finding its feet as a respectable subject. Its study over the past forty years has not yet produced a fully-fledged science, but it has produced a substantial body of knowledge. This book extracts from that knowledge many ideas and findings which are useful to our everyday lives.

This book will assist you to understand both the causes and the management of adverse events. It will enable you to take an informed view of risks which are not under your control, of accidents and disasters and wars, and of the competence and integrity, or otherwise, mostly otherwise, of the hierarchies responsible for their management. The book will help you to prevent or manage or avoid adverse events. It will be your ally, always on your side as you face the risks of life. It will help you boot them into touch.

THANKS

My thanks to Mrs Jo Hopperton who read my first attempt at the first chapters and gave very good advice.

My thanks, and probably yours, to Tony Shelbourn, highly qualified businessman, who read the early drafts of the chapters on decision-making and thinking, and very firmly instructed me to rewrite them.

My very warm thanks to my elder son Tobyn, who read everything several times and pointed out many faults, albeit with unnecessary relish.

My very warm thanks also to my British Airways colleague, Captain John Reed, alias Figment. His cartoons of corpulent Captains and sagging stewardesses have cheered the civil aviation scene for a generation. Figment is the embodiment of irreverence of authority and scorn for political correctness, the *Zeitgeist* of all who share his views. Unfortunately his disrespect extends to his old chums. As his sense of humour may offend those of a sensitive disposition, I have been obliged to remove several inappropriate inclusions from the cartoons in this book. I hope that I have not missed any.

INTRODUCTION

RISK: THIS IS YOUR LIFE

The management of risk is a down to earth subject. It is not an obscure science. Nor is it best left to experts. It is a universal activity. It involves us all. Everybody manages the risks of their life their every waking minute. If they manage them successfully, they are fed and healthy and happy at the end of the day. If they do not manage them successfully, then they are hungry or unhealthy or unhappy at the end of the day. That is if they are still here.

This book is a bridge between the theory of risk management and the solid facts of our everyday needs. It takes the concepts of specialists and applies them to real events, to well-investigated accidents. It explains the principles of risk management in plain language. It offers you the opportunity to cherry-pick, to take the morsels which you like and to leave the others for the birds. It will help you to understand and to manage the risks of life. It will help you to enjoy your life.

BACKGROUND TO THIS BOOK

I set out to write an academic book, a tome carefully observant of scholarly convention. A respectable publisher was kind enough to accept it. He planned to sell 800 copies at £50 each. I didn't like the £50. I buy many books, but none at that price. I wanted the book to be affordable to all. I then wrote an academic paper outlining some of my ideas. I presented it to an excellent friend, a fierce Professor of Risk Management, for his professional judgement. He is intellectually daunting. He is also ex-3 Para. Verbal combat with him therefore combines a mental workout with a degree of physical risk. He said that my paper would not be accepted by a leading academic journal because it contained terms which were not scientifically defined, such as common-sense. I made a jocular comment about academic unfamiliarity with common-sense from the far side of a stout oak table, sensibly keeping to myself my views on the common-sense of those who choose to jump out of fully serviceable aeroplanes.

By contrast, my late mother was a firm believer in common-sense. She had a double first from Cambridge. She was very bright indeed, even in her late nineties when I explained my ideas to her: ideas which I had taken many months to develop. She was a lady of few words; 'Sounds like common-sense to me, dear.' End of conversation. I knew that this laconic assessment was a compliment, but it was not clear that it equated to the award of a first *cum laude*. For one thing, she did not rise to her feet to congratulate me, perhaps because she was unwilling to disturb the cat. Rightly or wrongly I preferred her advice to that of the good Professor. I binned the projected academic book and set out to write a common-sense book on risk. And on the subject of parents, the poet Larkin was an ass.

Before I ended up in the academic world (loosely speaking), I was a pilot, first with the RAF and then with British Airways. In the 1980s and 1990s I was closely involved in national and international endeavours to advance aviation security standards. This gave me three valuable insights. The first was into preventing nasty things from happening – aeroplanes going bang. The second was into making good things happen – persuading HMG and other governments to get off their backsides and act. The third and most important insight was that it is not difficult to prevent a specific event (that is, a specific event only, not all possible events), but it is absolutely impossible to persuade bureaucratic governments into united and speedy action.

When I stopped flying I found that governments did not welcome consultants who told them that they were making a poor job of aviation security. I therefore transferred my attention to safety. I made a lengthy study of a particular bad verdict on an aircraft accident. This led to the development of a new method of analysing accidents, Failure Path® Analysis, an outline of which is included in this book. I found many good books on safety and on accident theory, but all written for specialists. I did not find a book which presented the interesting subject of risk management to general readers. So here we are. A book on how to understand nasty things, and how to prevent them from happening. Some theory, but no more than is helpful; and much practice, the bad to be vigorously opposed, the good to be warmly embraced. The central question of risk management is *how* to control risk. The answer is first to understand it. Start by reading this book.

PART I: WHY THINGS GO WRONG

Part I of the book, Chapters 1–4, looks at how, in life, things always go wrong. Actually they do not always go wrong, it just seems that way. We distinguish between Murphy's Law and Sod's Law. We find that Murphy's Law is not one law but two. This not only makes it the first law to prove itself right by proving itself wrong, but also the first law to prove that laws can prove themselves. Double first to Murphy. Murphy's Laws apply to human behaviour. Sod's Law relates to nasty coincidences, thunderbolts through our defences. Contrary to popular belief, there is no Law of Unexpected Consequences; failures of expectation are failures of our own making. Why toast lands butter side down is explained.

The management of risk is a contest between us and the nature of the world. The sage who said that we cannot command nature except by obeying her, omitted to say that to obey her one must first understand her. This book sets out to understand her. Life is a battlefield, so get stuck in. We can distinguish between general and specific risk. We can reduce some specific risks close to zero. We cannot reduce general risk to zero. The guided weapons with which we are attacked follow the paths of Murphy's two Laws and of Sod's Law. Our defences have a number of limitations, the gravest of which is complacency. Any course of action we take always comes with two inherent limitations; first, we forgo other options; second, our chosen course of action will have built-in disadvantages. Our attempts to reduce risk will introduce additional risks; this is one of the reasons why general risk can never be reduced to zero. Last, the world is dynamic. It is continuously changing and, worse, it changes in response to any action we take. This poses grave problems for our attempts at planning and control. Our prime objective should be the prevention of particularly nasty adverse events. Accident prevention is a continuous, unending process.

The meaning of cause, as in the cause of an accident, is not as simple as it seems. In fact it is a tricky question. There are various answers. This book suggests a new one: the cause of an accident is a failure of prevention. Any activity should be viewed as a system of operation. This accords with the contemporary conclusion that an accident is a system failure. Most system failures occur when a number of deficiencies lying dormant in the system become operative if they are suddenly connected together by some final event. This constructs a complete failure path through all the system defences. The result is an accident.

Think of a fruit machine. A row of fruit accumulates. We eagerly await the final piece which will complete the line to give us a jackpot row. In an operating system, we desperately do not want the final piece, because it will complete the line to give us a fatal row, an accident.

Any operating system has three objectives:

A. Successful conduct of the operation
B. Avoidance of extreme adverse events
C. Conformity with regulation

These objectives are not mutually supportive. They are in competition for time, for money and for all the other resources of the system.

PART II: THE LESSONS OF FAILURE

Part II of the book, Chapters 5–7, is concerned with the lessons which should be drawn from previous accidents and from accident investigation.

It starts by highlighting the problem of risk transfer. Those who decide on the acceptable level of risk rarely carry it: it is transferred to others. The risks of travel are transferred from the chairman of the airline or shipping line to their passengers; the risks of poor armour and unreliable weapons from the Whitehall warriors to our unfortunate servicemen; the risks of rotten pensions from a smugly index-linked chancellor to everybody outside privileged government circles. Everybody being greatly comforted by the privilege of paying for those government pensions.

Part II reviews a number of accidents. It draws general conclusions on the causes of accidents:

- With the possible exception of accidents arising from innovation, from failures of skill, and from the thoughtful acceptance of risk to oneself, all accidents are the product of faulty decision-making
- Most are failures of direction or management
- Verdicts of operator error should be taken with a pinch of salt Make that a bucket. In most cases there is a string of management failures in the fatal row
- With the possible exception of accidents arising from innovation, there is no such thing as an accident which could not have been prevented

The American research study of the shooting down of two US Army helicopters by the US Air Force in Iraq is a copybook investigation. The 1994 crash of the RAF Chinook ZD 576 on the Mull of Kintyre is a copybook example of the wrong way to bring a helicopter into service, and the wrong way to plan a flight operation; it is a copybook example of the wrong way to conduct an investigation, and of the wrong way to reach a verdict. Neither of these investigations reached satisfactory conclusions on the prevention of the accident.

PART III: THE PREVENTION OF FAILURE

Part III, Chapters 8–12, outlines the first two steps in the prevention of failure.

The first step is a new methodology of analysing accidents, Failure Path® Analysis (FPA). FPA conducts an objective audit of the operating system. It does not rely on the what-if hypotheses of fault trees. It asks the question: what would have prevented the accident? Its unit of analysis is the PO2PL, Possible Opportunity to Prevent Loss. It locates the failure points in the system and traces the accident path through it. It identifies the points of responsibility. FPA does not address issues of blame. That is a more subjective question; its investigation is a different, quasi-judicial, process. FPA turns the art of accident investigation into a science of accident prevention. The application of FPA to the US Air Force shoot-down of two US helicopters reveals the details of multiple failures. The application of FPA to the crash of RAF Chinook ZD 576 reveals an extraordinary catalogue of administrative and operational system failures in the RAF.

The second step towards accident prevention is sound decision-making. The short Chapters 9–12 deconstruct the decision-making process. Decision-making is as much art as science. A sound decision balances on a three-legged stool, on information, on intelligence and on integrity. If any one leg fails, the decision fails. Unfortunately it is not usually the decision-maker who falls flat; it is those who are relying on him. We find that decision-making is not a rational process. It is judgemental. The better we understand the process, the better we can make it serve our purposes.

PART IV: SAFETY IN SIGHT

Part IV, Chapters 13–15, takes our three final steps towards the prevention of failure.

Chapter 13 advocates joined-up thinking, the careful balancing of system objectives with their risks, and a system of operation which incorporates active defences against risk. It advocates the pursuit of security as a model for the pursuit of safety. In security, the antagonist is an evil-doer who wishes to do us injury. In safety, the antagonist is the mischievous nature of the world which seeks to confound our activities. The principles and processes of actively defending ourselves are much the same in each case.

Chapter 14 examines the risks over which we have no direct control, those to which we are subjected by hierarchies. These risks range from avoidable wars to the diversion of precious resources into the micro-management of utter trivia. The chap who said *de minimis non curat lex*, the law does not care for trivia, was unacquainted with our Mongol hordes of regulators, the nannying nincompoops who beset every corner of modern life. Lucky man. We look at a possible strategy of safe navigation through the dilemma of too little control or of too much control.

Chapter 15 concludes that good information is the key to our ability to assure our own safety. We want the information which is good for us. The hierarchies prefer us to have the information which is good for them. There is a long war to be fought in the interests of truth.

With good information and good decision-making we will be on track to join up our thinking. We know from this book that our enemy, the mischievous nature of the world, will observe any gaps in our thinking. Gaps in our thinking translate into gaps in our defences: ready-made entrances for thunderbolts. Read this book and close as many as you can.

SAFETY LESSONS FROM AVIATION

This book draws many lessons from aviation, for two reasons.

First, I am reasonably qualified to sort truth from lies in aviation, I certainly do not claim the same in other areas.

Second, a powerful case can be made that the best safety standards in the world are to be found in (the best of) aircraft operating standards. As we will see later in the book, aviation safety practice is at least fifty years ahead of NHS safety practice. How do I know that? Because I started flying fifty years ago, and to this day the NHS does not do what we did then. Moreover, in general, major aircraft accidents are investigated by true experts. Their reports are little distorted by outside interests, and they are publicly available. The lessons of aviation are accessible or easily learned.

Aviation is not a dangerous business for those who know and respect its rules. For those who do not know its rules, or who know them yet scorn them, retribution awaits. A sharp reminder will be the very last thing to pass through their minds. It is this immediacy and finality of retribution, and particularly its cost, which drives aviation safety. Paradoxically, danger makes for safety, and safety makes for complacency and for danger.

For these reasons this book draws much on aviation in its discussion of good general safety practice.

THE ACID TEST

This book seeks to be practical. It seeks to be relevant to real life. It draws from real-life for most of its case studies and its examples. Subject its suggestions and models to a real-life test, that of your own experience and your own observations. Try to prove it wrong. I hope that you will not easily succeed.

REFERENCES

It is customary in books which cite other books and papers to provide, in a footnote, a reference to each source. This enables the reader to check the accuracy of the quotation and to examine its context. Unfortunately such footnotes have a pernicious effect on readability; they break up the flow of the reader's thought and they make the book look a mess. Therefore I have gathered all references into endnotes at the back of the book.

Enjoy the book.

INTRODUCTION TO PART I
WHY THINGS GO WRONG

CHAPTERS 1–4

Most of us have a general impression that things always go wrong. In Part I we examine whether this is true, and we find that nature's default mode is indeed to oppose us. We seek to understand why, and we see that a classical model of the world offers a useful explanation.

In Chapter 1 we investigate how things go wrong; we find that our undoing lies in the ubiquity of risk, and in the workings of Murphy's and of Sod's Laws. We identify and explore the important difference between these laws. We see that they are Laws of Nature, so there is no avoiding them. But we look on the bright side. The better we understand the laws of nature, the better we can deal with them. And deal with them we can.

In Chapter 2 we examine the battleground. We find that some challenges arise from human fallibility, and others arise from the cussedness of nature. This enables us to begin to develop the Rules of Battle. By the end of Chapter 2 we have the whole framework of our battle with the Gods.

In Chapter 3 we dip our toes into the waters of risk management. We find that they are navigable, if not entirely fathomable. We distinguish between general and specific risk. We formulate a Revisionist Normal Accident Theory. It states that the possibility of accidents is normal in all operating systems.

In Chapter 4 we look hard at the meaning of accident cause, and even harder at some inaccurate but convenient interpretations of cause. We find that any accident or loss is a failure of our system defences. Such failures occur when some final event runs through a system, connecting the accumulation of small cracks in the system into a single, total fracture. The final event is the lightning strike by the Gods which we thought, or hoped, would not happen. We call the components of this fracture a fatal row.

To avoid extreme adverse events we must avoid being zapped by the Gods. To have any chance of this, we must always remember that the Gods are always waiting patiently for us to forget that they are always waiting patiently.

CHAPTER 1

WHY THINGS GO WRONG

NATURE'S LAW OF PARADOX

Nature's Law of Paradox integrates risk with life, with activity and with assets. For as long as you live, you are at risk of death. When you are dead, you are no longer at risk. But you are dead. For as long as you indulge in an activity, you run the risks of that activity. Rugby players risk injury; motorists risk motor accident; adulterers risk rolling-pins. For as long as you own an asset, you face the risk of losing it. Your brain may slow down, your house may burn down, your car may be stolen, your spouse may be seduced. Asset risk ends only when possession ends. Activity risk ends only when activity ends. Life is risk and risk is life. Eliminate the one and you eliminate the other. The management of life is the management of its risks. That is what this book is about.

WHAT IS GOING ON?

How-to books are spread along the dreamers–doers continuum, a spectrum from philosophy (how to think) at one end to manuals (how to do it) at the other. This book sits in the middle; it is a how-to-think-how-to-do-it book, a book for thinker-doers. Its central concern is to clarify the paradox binding risk with life. Our existence is governed by the laws of nature; the law of paradox is one of them.

We cannot change the laws of nature. Our ancestors in the trees needed to make sense of their world in order to survive from one day to the next. We in the modern Western world are more privileged.

Our day-to-day survival is not generally an issue. To a large extent we are able to concern ourselves with extracting the most from our existence. For this, we too need to make sense of our world. We live very different lives from those of our ancestors, but we live in the same world, ruled by the same laws of nature. Nature does not explain its laws. It is up to us to induce them from experience. The better we understand them, the better we can manage them. The question most asked since the beginning of time is 'what's going on?'

PHILOSOPHY – LITE

Let us make a brief excursion to, and then a rapid exit from, the deep waters at the philosophic end of the thinking–doing spectrum to ask: why is the nature of things the way it is? The religious would reply that it is God's will and that is the way he wants it. By postulating a God or Gods the religious construct a belief system which makes sense of the world, if only to those who choose to believe it. Those of a more doubting disposition might see the concept of God or Gods as the direct product of our psychological need to make sense of the world.

The doubters also wish to make sense of the world. They start by suppressing their tendency to find a cause for each effect. They do not seek any concept or construct outside nature to answer why nature is the way it is. Their answer is because. That is, because that is the way it is. Nature is all-powerful, powerful enough to originate itself and to determine its own laws. Which is much the description that the religious would give to their God. Therefore the doubters' laws of nature are agreeably synonymous with the believers' laws of God. To manage our existence we need to work out what those laws are. We do not need to know who wrote them.

THE GODS – LITE

The human need to make sense of the world is allied to a hard-wired tendency to explain every effect by finding or fabricating a cause for it. Our ancestors' survival was dependent on their continuous success in defending against life's risks, that is against the slings and arrows of outrageous fortune. It is therefore not surprising that a particularly thoughtful group of ancients postulated the existence of a set of highly temperamental Gods, and with a nasty vindictive streak. To some, the ancients' model of a bunch of boozers on Mount Olympus makes more sense of the vicissitudes and horrors of life than does a model of an all-loving God, or of a benign and beautiful Nature. Their construct saw us as the playthings of the Gods, the ebb and flow of our lives being entirely ruled by their laws. They do what they wish with us. What we can usefully do is to identify the rules of play.

THE GODS' RULES – OK

Rule One The Gods are omnipresent. Trouble-makers are always hanging around the street corners of life. Disaster is always lurking.

Rule Two The Gods have a great sense of humour. Unfortunately it is generally malicious. They relish any opportunity to have a laugh at your expense.

Rule Three Try hard not to attract their attention. Do not act the goat, that is do not take stupid or unnecessary risks: they will undoubtedly notice. Remember *quem deus perdere vult, prius dementat* (whom the Gods wish to destroy, they first make mad). Therefore think; do not take thoughtless risks. And when you take necessary risks, be thoughtful. Calculate and minimise the downside, the duration and the extent of the risk. You will thereby minimise your exposure to the Gods' attention. If you do attract the attention of trouble-makers, they will trip you up. That is what trouble-makers and Gods do.

Rule Four This is the most important of all rules. **Do Not Ever Provoke The Gods.** Do not think – and whatever you do, do not say – it cannot happen to me. That is *hubris.*[1] Nothing annoys the Gods more. They will be furious. Do not kid yourself that they are not there, do not think that you can handle them, do not award yourself membership of their gang, never be complacent. Never ever think that you have got it made. All such arrogance is the equivalent of blowing raspberries at trouble-makers. You are asking for a kicking. You will get it. The Gods will launch a thunderbolt. It is their purpose-built, hubris-seeking missile.

Rule Five If despite your best and most thoughtful behaviour, the Gods trip you up or knock you about, take it like a man. Do not say these things happen, but why me? It happened because the Gods wished it to. It may have happened because your risk management failed, in some way you broke a rule and annoyed the Gods. They will be expecting you to remedy your error. If you do not, you will seriously provoke them. They will scent *hubris.* They will alert a stand-by thunderbolt. Therefore revisit your system of management before the Gods revisit you. Be honest, identify its true points of failure and put them right. If you will not do this or, worse, if you are gutless and blame a subordinate or somebody else and think that you have got away with it, you may fool some people, but you will not fool the Gods. Sooner or later they will give you a kicking. Your subordinates and those people whom you have not fooled will be praying that it will be soon and on earth, so that they can watch. And join in.

1 Hubris is the Greek ancients' word for the belief that you are invulnerable, that no misfortune can befall you. Bad mistake. Hubris is a very useful word and continually reappears in this book.

THE IMPORTANT DIFFERENCE BETWEEN NICE AND NASTY

This book is about the avoidance of unpleasant events, things we do not want. Therefore it uses the word 'risk' only to refer to something unpleasant. This usage deliberately excludes happy accidents – the reaping of unexpected benefits or the hitting of unforeseen jackpots – from our concept of risk. In general and academic usage, risk may be good or bad; this is singularly unhelpful because life is all about distinguishing the good from the bad. The rules for obtaining nice things are not the same as the rules for avoiding nasty things. Interestingly, as we will see, the rules for obtaining nice things include, and indeed most likely start with the rules for avoiding nasty things. Nonetheless the two sets of rules are very different. The downside of using the word 'risk' exclusively as a pejorative is that it may well cause academic brows to furrow and academic fingers to wag. That is a risk which we must take.

MURPHY'S LAWS

Murphy's Law has been commandeered by popular usage as the title for any aphorism expressing the general cussedness of life, the feeling that things always go wrong. But it is not true that things always go wrong, it just feels that way. It is true to say that things always *can* go wrong. And now we are on the right track.

In earlier happier days Londoners could suffer congestion without being charged for the experience. They thought that Murphy referred to the firm of builders who appeared to specialise in digging up busy roads and then disappearing, leaving only their name-boards, Cheshire Cats grinning in triumphant valediction. In fact the builders' claim to fame should be subordinated to that of Captain Ed Murphy, a pilot, serving after the war in US Air Force development projects. Exasperated by the cack-handedness of one of his men, he famously said in so many words: 'if there is a way of assembling something wrong that son-of-a-bitch will find it'. It is perhaps an injustice that the law carries the fair name of Murphy, and not the name of the serial blunderer. On the other hand, it is a pleasing (if unfair) reinforcement of the stereotype of an Irishman. Captain Ed Murphy's statement of his law encapsulates an important truth: if something can be done wrong then, sooner or later, somebody will do it wrong. Note carefully that it relates to human action; it requires human agency.[2]

The difference between Murphy's Law and Sod's Law has been the subject of much dispute. In 2001 the literary editor of *The Times* described a series of misfortunes which he ascribed to Murphy's Fifth Law, regrettably and reprehensively omitting to inform us of Murphy's first four laws. The correspondence columns of the paper are usually to be relied upon for *ex cathedra*[3] pronouncement on important matters. On this occasion, the letters gave amusement; unfortunately (like politicians), when laid end to end they did not reach a useful conclusion. An early correspondent supported the broader version of Murphy's Law given above: if things can go wrong, they will. He affirmed that Sod's Law states that if there are two possible conclusions, the least favourable will occur. Father Bryan Storey agreed with this version of Murphy's Law, but with the important proviso that when things go wrong, it will be at the worst possible time. Professor Ian Fells, Fellow of the Royal Society of Engineering, cited his personal research into Murphy's Law. He had tracked down the eponymous Captain Murphy, still alive and well in the US. But when the professor went to interview him, he found that Captain Murphy had Alzheimer's and could remember nothing.[4]

As Murphy's and Sod's Law are widely held to be the causes of many accidents, it would be useful if we could identify the conceptual difference between them. And we can. Murphy's Law addresses

2 Should we campaign to rename the law? There are many with much better claims to the Olympic gold in serial blundering. Bush's Law? Blair's? *Embarras du choix.* On balance no simple law could adequately cover the past decade. We need a big book, *The Bush-Blair-Blunder-Book of How to Arrange Avoidable Disasters and Kill Lots of People and Make Enemies Everywhere.*

3 *Ex cathedra*: lofty, superior, toffee-nosed, ie, we know better than you.

4 By some definitions this could be deemed the ultimate, incontrovertible proof of Murphy's Law. However, Professor Fells judged that Murphy's Law was inadequate to explain so piquant a misfortune. He postulated a superior law, O'Reilly's Law. It states that Murphy's Law is wildly optimistic.

human performance. Sod's Law addresses the workings of nature. Here's why. You will remember that the Captain Ed Murphy derived his law from his observation of one of his men. This unfortunate, given an opportunity, would assemble aircraft parts incorrectly. In other words, if a part would fit the wrong way round, or in the wrong place, that is how he would fit it. Captain Murphy's grief was caused by human behaviour and in particular that element often called, fairly or unfairly, human error. Murphy's beef was about man, not God. As we will see in a moment, Sod's Law is not about man, it is about God and the nature of the world.

But Murphy's Law is a general law. It not restricted to human behaviour, it applies equally to human systems of operation.[5] Arguably, a variant of Murphy also applies to natural systems of operation, but that is a separate subject.

Murphy's Law applied to an action we will call Murphy's First Law.

MURPHY'S FIRST LAW

Murphy's First Law relates to an action. It states:

If something can be done wrong then, sooner or later, somebody will do it wrong.

If a washing machine can be loaded and started without the washing powder or liquid, then sooner or later someone will. And it can. And I have.

If you can leave your house without your keys, and close the self-locking door, then sooner or later you will. I haven't. Yet.

Murphy's Law applied to an activity or system of operation we will call Murphy's Second Law. Should the pernickety demand the

5 Sharp intake of breath from my academic colleagues scenting a possible clash with the conventional definition of system. Breathe on, Ladies and Gentlemen, your definition is fine but just not helpful to this discussion of everyday risk.

evidence for the assertion of Murphy's Second Law, let them consult universal experience.[6]

MURPHY'S SECOND LAW

Murphy's Second Law relates to activity. It states:

If error can be introduced into a system of operation, it will be.

If hospitals can inject the wrong drug and kill a patient, they will. They have, and will again.
If military aircraft can attack friendly forces, they will. They have, and will again.

Systems are purposeful. Unlike people, they are purpose-built to achieve an objective. In the case of hazardous undertakings: aircraft operation, nuclear power generation, declaring war, etc., systems are purpose-built not to go wrong. But they do. And when they do, it means that the system has failed. The system designers or managers or operators have failed. The managers' first recourse will be to blame the unfortunate operator – pilot error again (sub-text: not our fault). Don't be in a hurry to believe them. Their second recourse is to present the adverse event as a circumstance beyond their control. Clearly no failure of the system for which they were responsible. Further lies to follow.

In our lives and activities, we endeavour to persuade things to go right. The Gods enjoy arranging for things to go wrong. They have three primary avenues of attack. Murphy's First Law and Murphy's

6 Second sharp intake. It is unacceptable in an academic work to support an assertion by citing universal experience or common-sense. It is acceptable to cite one person if he/she is an acknowledged authority, preferably the book reviewer if the author knows what is good for him. We may wonder why a lesson by an academic is to be preferred to the universal lessons of life. Anyway this book does not purport to be academic. It aspires only to be good, or common, sense.

Second Law are their first two. Sod's Law is their third. We will discuss their secondary avenues of attack in Chapter 2. Let us now look at Sod's Law.

SOD'S LAW: WHY THINGS ALWAYS HAPPEN AT THE WORST POSSIBLE TIME

If, in a particular activity, there is a 1 in 1000 chance of a certain untoward event then, if we undertake the activity 1000 times, on one occasion the untoward event will occur. (There are various statistical ifs and buts which should attach to this statement. They would be an unnecessary complication, so we will tip our hat in their direction but otherwise ignore them.) What we do not know, and what no statistics can tell us for certain, and wouldn't it be wonderful if they could, is on which of the 1000 occasions disaster will strike. In a pure case in which each occasion is totally independent of all other occasions and is subject to absolutely no other influences, then the likelihood of the untoward event is the same for each occasion. Its occurrence is subject to the rules of chance[7], a subset of the laws of nature. Nature chooses the time of its occurrence. It chooses the worst possible time if it so wishes. And it usually does. The Gods have a great sense of humour. They find your discomfort greatly entertaining.

We can respectfully disagree with Father Bryan Storey's submission to *The Times*; it is not an important proviso of Murphy's Law that, when things go wrong it will be at the worst possible time. Murphy's Law relates to (what we will call for the moment) human error. The timing of that human error is in the hands of the mischievous Gods. It is ruled by an entirely different law, a law of nature, Sod's Law.

SOD'S LAW: BUTTER SIDE DOWN

But why does toast always land buttered side down? Obviously toast does not *always* land buttered side down, it just seems that way, and to

7 Stochastic is the heavy word for this, for those who like heavy words. For those who don't, that's what it means.

a lot of people. The contrariness of life is a topic of universal interest. If you type 'butter side down' into Google, you will get 7,600,000 hits. 'Why do things always go wrong?' is a common cry, but they don't always go wrong, just rather too often, and usually at the most inconvenient time. The BBC's guide to life correctly identifies the height from which the toast is dropped as a factor. However, it also states that the uneven mass distribution of the butter is a second factor, but without citing either the source or the scientific grounds for this assertion. The BBC's guide to the performance of toast would be rejected out of hand by any academic referee. Remember that next time the BBC comments favourably on the performance of Her Majesty's Government (1997–2008).

However, the objectivity of at least one learned paper on the landing performance of toast has been approved by academic referees. Robert Matthews' *Tumbling Toast, Murphy's Law and the Fundamental Constants,* published in a respectable academic journal, contains an altogether more thoughtful and scientific examination of the question. Robert Matthews rejects the BBC's assertion that butter affects the aerodynamics of toast. He shows that the key factors are the convention of buttering the top side, together with the height of the table from which it drops.

His analysis then becomes fascinating. Tables, he says, are the height they are because human beings are the height they are. His research shows that the maximum safe height for human beings is around 3 metres. Above that height a simple fall would accelerate the skull to the point at which the chemical bonds in the skull would rupture, causing severe fracturing. Our normal height dictates that our tables are not high enough to allow toast to rotate through 360 degrees and land the right way up. But then, writes Robert Matthews,

> ... comes the cosmic twist in the tale. The formula giving the maximum height of humans turns out to contain three so-called 'fundamental constants of the universe'. The precise values of these fundamental constants were built into the very design of the

> universe just moments after the Big Bang. In other words, toast falling off the breakfast table lands butter-side down because the universe is made that way.

Robert Matthews' proof is elegant. Note his conclusion: 'because the universe is made that way'. In other words, descending toast obeys the laws of nature – you would hardly expect nature to grant it an exemption. Very obvious – after Robert Matthews has explained it to us. We need not look beyond those laws for the explanation of butter-side down.

One obvious simple response is to butter the other side of the toast. Apart from acknowledging the cosmic laws of the universe, it would also stick the toast to the plate. Problem solved. But new problems are created. Butter wasted, butter adhering to the plate lost to the breakfaster, washer-upper displeased. And what about the marmalade? Marmalade the top and invite marmalade-side down arrival on the carpet. Marmalade the bottom and the marmalade will target your tie en-route from your plate to its fate. The solution which sounds simple turns out, as so often, to be simplistic. Any solution to any problem brings with it its own new problems. Cosmic laws cannot be beaten and they cannot be avoided, not even by toast.

SOD'S LAW: A LAW OF NATURE

If we deconstruct the fall of the toast, we can identify two separate phases. The first is the phase culminating in the release of the toast; the second is the phase from the release of the toast to its arrival on the carpet. We have been discussing only the second phase, the performance of the toast on its descent. During this phase there is no human intervention (we are assuming a clean release and not a directional launch). Robert Matthews is therefore right, the flight of the toast is dictated solely by the laws of the universe. And that is the essence of what we call Sod's Law – it is a manifestation of the workings of nature. Murphy's Laws needs a human input. Sod's Law does not. Nature can manage without us.

SOD'S LAW: A LAW OF COINCIDENCES

Coincidence, that is unfortunate coincidence, is central to many untoward events, whether trivial or serious: why was it on the day after the new carpet was laid that you dropped the toast? Why was it on the one day that you were late out of the house that there was a speed-trap? Why was it on the day that all the flight-crew of an arriving British Airways 747 were ill that there was thick fog at Heathrow?

Coincidence may be of time or of place, mostly it is both. The toast was dropped at the time immediately after the new carpet was laid. Also, it was dropped at the place where the new carpet was laid. When two cars collide on a bend, they both arrive at the same time at the same place. Had it not been *at the same time* that they rounded the bend, no collision would have occurred. Had they not been *in the same place* at that time, rounding the same bend, no collision would have occurred. There is a whole methodology of risk management which is based on this. It is called a barrier methodology. It focuses on putting barriers of time and/or space between events the coincidence of which is to be avoided. In its simplest form it is very useful. Keep the toddler out of the kitchen *where* pans are boiling – a *place* barrier. Do not smoke *when* refuelling my aircraft – a *time* barrier. In its complicated form it is complicated.

Just in the wrong place at the wrong time is a necessary co-incidence. It is not necessarily a coincidence. Many apparent amazing coincidences turn out – after a little careful thought – to be less than amazing. Some are almost predictable. Your wife was absolutely determined that, on the day after it was laid, no one would drop a crumb on her new carpet. At breakfast she treats each member of the family to her best death stare. The children are completely unperturbed; they munch on. You feel a familiar tightening of the stomach. What have I done now? (Nothing. Yet.) You tense up. Your hand tightens on the toast. Ping – off it goes. No coincidence, just a simple sequence of events: new carpet – uptight wife – death stare – increased anxiety – muscles tense –

motor performance of hand disrupted – toast launched. Not coincidence but cause and effect. The arrival of the new carpet has increased the probability of a mishap. Not an act of nature, not Sod's Law. Not actually gross negligence on your part, but don't bank on the Board of Inquiry finding herself at fault. More about Ping below.[8]

MURPHY'S LAW AND SOD'S LAW OFTEN ACT TOGETHER

Many accidents involve the operation of Murphy's Law and of Sod's Law. When motorists collide on a bend, Murphy's Law explains that one or both were probably guilty of human error, Sod's Law dictates the coincidence of time and place. Your release of the toast was a human failing, albeit very understandable. Toast can be dropped; sooner or later somebody will drop it. Murphy's Law. It was you, unwitting victim of cause and effect, not Sod's Law. Its flight path and butter-side down landing were Sod's Law. Not Murphy's Law, not you.

Allegedly, Sod's Law was once proven in the following manner:

> It is said that a professor was lecturing to his students. 'You all know,' he told them, 'that when a piece of toast is dropped, it invariably falls buttery side down. That is called Sod's Law. Now observe.' He took a piece of toast, buttered it, held it out and dropped it. It fell buttery side up. 'There you are', he said, 'a perfect demonstration of Sod's Law'.

Thus a law of coincidence was proven by a coincidence. We can only admire the Gods for the breathtaking brilliance of their eye for detail and for their unsurpassable sense of humour. Actually the

8 If you have ever dismantled a mechanism containing several small springs you will know that, despite your very best efforts, one of them will get away from you. In an instant dash for freedom, it will ping across the room. The technical term for this is Ping-Fuck. If the spring disappears through the only crack in the floorboards, you may plead Sod's Law. If there are lots of cracks in the floorboards, it is you, and serves you right. You provoked the Gods.

professor also did well; he was on a winner whichever way the toast landed.

The most complex accidents will almost always involve coincidence. To breach system defences, a set of highly complicated events has to occur, and most probably coincide with another set of events; both sets must circumvent the best risk management procedures that man can devise. Unlikelihood upon unlikelihood. But it happens because it can happen. And if it can happen, it may happen. And if it

CASE STUDY 1.1 NICK'S MOBILE PHONE

Murphy's Law and Sod's Law Act Together

Nick was my personal trainer. He was in his mid-twenties, a bouncy rubber ball of muscular sportsman. Footballer, boxer, bon viveur, occasional trouble-maker, he was a player, determined to make the most of life. Intensely gregarious, he rushed about between his personal training clients, his aerobics classes, his many friends, and his troublesome wife (later ex-wife). He was continually on his mobile phone, often running late and rearranging his appointments, firming up the attendees for the evening's party, sometimes apologising for the occasional no-show, not always unrelated to the previous evening's party. He was of the generation which appears to have a mobile phone surgically implanted.

Nick was an excellent personal trainer. He was unmerciful. He made no apparent allowance for my age (not your business, but over sixty). Any suggestion that a weight was too heavy was ignored; my attempts at kick-boxing were derided (my Granny can kick harder than that. If I were your Granny, I'd kick harder than that.) He was always fun and my fitness benefited greatly. I also found that 10 minutes of cardio-vascular workout is a much better wake-up call to the brain than any number of cups of strong coffee.

He came to my house regularly for two and a half years, a total of over 200 visits. He drove the 12 miles of country roads in his 8-year-old bright red BMW. Uniquely on one visit, perhaps his 180th, he forgot his mobile phone. It was a rotten, rainy day. And almost certainly you have guessed what is coming next. Soon after he left me for his next client he had a big puncture, a total flat, when he was deep in the countryside. He had no spare. He had to walk 2 miles in heavy rain to find help.

His forgetting of his phone and his failure to carry a spare wheel were the workings of Murphy's Law.

It was a coincidence that the one day of his puncture was the same one day that he had no mobile phone. And coincidences are the working of Sod's Law.

cannot happen, it may happen. And if we are sure that it cannot happen and become complacent, we will provoke the Gods – breach of Rule 4: Don't dis the Gods. They will be very upset (cause). And then it will certainly happen (effect). Cause and effect. Self-inflicted injury. You ask for it, you get it. That's the way it works.

ACCIDENTS ARE NORMAL

It is common knowledge and sadly common experience that accidents can always happen. For reasons which will become progressively clearer chapter by chapter, our endeavours to avoid all accidents are endeavours to beat the Laws of Nature. And however hard we try, we can never do that. In a simple domestic setting things unfailingly go wrong, even in the best run household. Keys are lost. Toast is dropped. Vases go flying. Precious possessions are broken. People slip. People trip. Factor in either inexperienced toddlers, whirling-dervish children, dopey youths, truculent teenagers, the lovelorn, the infirm, or the elderly, and the setting is more accident-prone. If it is a badly run, dysfunctional household then it is accident-certain, so stand well back as it ricochets from one disaster to the next.

Exactly the same is true of organisations. Some are well-run; some are not. Machines break. Computers misbehave. Software acts up. Deliveries are late. People are late, hasty, indecisive, unreliable, incompetent. Each person is a source of human error. The more people involved, the more the sources of human error. As organisations and processes grow in complexity they fall victim to a double-whammy. Increasing complexity not only increases the number of possible sources of accidents, it also increases the opacity of the organisation. It makes it more difficult to understand all that is going on, not to mention all that might happen. The best navies in the world have accidents. The best airlines in the world have accidents (the worst have lots more). Nuclear power stations go bang. Organisations can no more eliminate all accidents than people and families can. In all situations accidents are normal.

BOOKS OF RULES: OK? NO

Nor do instruction books, books of rules or libraries of procedures provide the answer. In modern life there is a book of rules or code of practice for just about everything: building regulations, planning regulations, health and safety regulations, sweeping chimneys, selling houses. Who reads the instruction books for telephones, videos or motorcars? Not many of us. Do not ask how many understand them. A book of rules for the simplest operation rapidly becomes very long, well before it covers each and every eventuality. However long a book of operational procedures becomes, it is certain not to cover every eventuality.[9] And the bigger it grows, the more difficult to use it becomes, so it falls into disuse. Just as military plans do not survive contact with the enemy, books of procedures do not survive contact with real life. The world is in a state of flux, situations change and rule books cannot keep up. Operators at the sharp end have to improvise

9 Malik's Third Rule of Regulation. See Chapter 15, Appendix.

and make their own rules if the operation is to continue. They are also likely to make their own rules in order to make their lives easier. One way or another, the custom and practice of the real world drifts away from the paper strictures of the distant world in which the regulators live their comfortable, index-linked pensioned lives. It is one example of what, in our next chapter, we call 'shifting sands'. It is a Law of Nature.

NATURE RULES: OK? YES

To the Laws of Nature we are toast. They do not depend upon us for their existence. They work just as well in our absence. If we are present they work on us. If toast is present, they work on toast. *Pace*[10] those existentialists who propound that if a tree falls in a forest, it has not fallen until a human being perceives its fall. Jumping from the reasonable proposition that our perception of an event is essential to our awareness of its occurrence, they propose that our awareness of the event is therefore essential to its occurrence. The logical extension of this is that a danger is not a danger until someone perceives it a danger. So it is OK to drive down a motorway the wrong way until someone notices. The laws of nature do not apply to us until they hit us. Brilliant. This belief must be an unsurpassable benchmark both of self-importance and of stupidity. *Les Deux Magots*[11] seem to have lost a 'g' somewhere.

10 *Pace* (Latin) means that I am politely expressing a different opinion, not trying to start a fight.

11 *Les Deux Magots* is a pleasant little coffee-house and restaurant in the artists' quarter of Paris. Deux means two; *Magot* has different meanings. *Les Deux Magots* is famous as the meeting place of existentialists Jean-Paul Sartre and Simone De Beauvoir, two of the most overrated of France's overrated philosophers. Sartre is widely held to have been the inspiration of Pol Pot, the architect of the genocide in Cambodia in which 2 million were murdered. So although *Pace* (footnote 10) normally means I politely disagree, in this case of disagreement with two self-indulgent monsters of self-delusion, Sartre and De Beauvoir, *Pace* means Balls to Both of You.

SUMMARY OF WHY THINGS GO WRONG

1 Nature's Law of Paradox states that risk is integral with life, with activity and with assets. In other words, risk is part of the package called life
2 Asset Risk is inherent in the possession of any asset
3 Activity Risk is inherent in the undertaking of any activity
4 If our lives are to be safe, prosperous, happy and fulfilling, we must manage all the accompanying risks
5 To manage anything, we must first make sense of it. The world does not provide us with a book of its rules; we must induce how it works from our experience of it. Wherever possible we should learn from the experiences of others, because many experiences are painful; some are fatal
6 Our common purpose is to understand the rules. We have no need to agree who wrote them. We will refer to them as the laws of nature. Some will believe them to be the laws of nature. Some will believe them to be the laws of (their) God. Some will believe them to be the laws of the Gods
7 Our subject is risk. We suffer continual bombardment by the slings and arrows of outrageous fortune. We need to be active in defending ourselves. The attacks that we face appear to be arbitrary in their targeting, their methods and their timing. We must make sense of them if we are to defend ourselves successfully. The ancients saw their attackers as the Gods, the celestial masters of hit, run and laugh attacks. The Gods' targets of choice are us, their playthings. Life is guerrilla warfare with the Gods. The ancients' model makes temptingly good sense of the incoming slings, etc. of outrageous fortune
8 Our subject is the avoidance of unwanted events. Therefore we define risk as the risk of nasty things
9 Murphy's Laws are laws of nature which apply to human behaviour. They require a human agent

10 Murphy does not confine his interest to human actions. He extends it to all human endeavour. Murphy is the Kilroy of risk. Murphy is everywhere[12]

11 Murphy's First Law states that if something can be done wrong then, sooner or later, someone will do it wrong

12 Murphy's Second Law states if error can be introduced into a system of operation, it will be

13 Sod's Law is a law of nature which applies to the working of nature. It does not require a human agent. It does enjoy a human victim

14 Sod's Law is the chance coincidence of two or more events or conditions. The coincidence may be of time or of place; most often it is of both

15 If something happens at the worst possible time/place, we call it Sod's Law. If it happens at a propitious time/place, we call it good luck. This is unfair on Sod

16 Many accidents involve the combined operation of both of Murphy's Law and of Sod's Law. The occurrence of the human 'error' is Murphy's Law. The operation of Murphy's Law at a bad time and/or a bad place is Sod's Law

17 Accidents are normal. It is a law of nature that they can never be totally eliminated

18 Books of rules can never succeed in eliminating all accidents. Some books of rules help. Lots don't

19 The laws of nature work fine without us. If we are around, they work on us. If you don't believe that, hang about, thunderbolt coming

20 Chapters on defensive planning later. Read on if you wish. Otherwise thank you for reading thus far

21 Always be ready to duck

12 I laugh and shout and jump for joy, for I was here before Kilroy.
– Sorry to spoil your little joke. Kilroy was here, but his pencil broke.

CHAPTER 2

MORE ABOUT WHY THINGS GO WRONG

SOURCES OF DIFFICULTY AND DISASTER

Self-Inflicted Injuries are our most certain route to disaster. Self-inflicted injuries are the product of human fallibility. Human fallibility is a fact of nature. It cannot be eliminated. It can be minimised. For that reason we spell out the rules of battle. If we carefully adhere to them, we minimise the opportunity for human fallibility to sabotage our endeavours. The avoidance of self-inflicted injury is the object of the first four rules of battle. Here are the first two:

- First rule: know the rules of battle
- Second rule: observe the rules of battle.

Human Fallibility is reflected in *Murphy's two laws*, in *insufficient thinking*, and in *complacency* and *hubris*. Hubris is too terrible to contemplate; it is the belief that the Gods cannot touch you, that the rules do not apply to you. Hubris triggers thunderbolts. Complacency is the belief that the Gods will not touch you, that you can get away with breaking the rules. Insufficient thinking is the failure to identify all the factors crucial to a situation, and their impact on its outcome. And it is obvious, is it not, that the rules of battle are the first of the factors crucial to a situation. Therefore any breach of those rules is proof of insufficient thinking. An obvious mistake and culpable. Let us not make it. And we will keep a watchful eye out for Murphy.

Nature's Little Ways is a polite term for the cussedness of nature. It is at the heart of Nature's Law of Paradox, the law which states that risk is inherent in life.

Sod's Law, our old friend, is top of the list.

Options are Forgone. When we decide upon an objective, we forgo the options of other objectives. When we decide upon an activity, we forgo the options of other activities.

Built-In-Buts are the disadvantages which accompany any activity.

Go by bike. Advantages: cheap, healthy. *Buts*: slow, cold, dangerous. Travel by car. Advantages: fast, warm, safer. *Buts*: costly, sedentary.

Double Agents are our own risk reduction measures. They introduce their own risks, as we will see shortly. To some degree they work against us.

Shifting Sands is the name which we give to the changes in the situation in which we are working.

These are the forces arraigned against us. Whence the next two rules of battle:

- Third rule: understand the forces arraigned against us
- Fourth rule: deal with the forces arraigned against us

Ignorance or breach of any rule is a self-inflicted injury.

THE FORCES ARRAIGNED AGAINST US
THEIR ORDER OF BATTLE

SELF-INFLICTED INJURIES

Ignorance of the Rules of Battle
Breach of the Rules of Battle
Not Understanding the Forces Arraigned Against Us
Not Dealing with the Forces Arraigned Against Us

HUMAN FALLIBILITY

Murphy's First Law
Murphy's Second Law
Insufficient Thinking
Complacency
Hubris

NATURE'S LITTLE WAYS

Sod's Law
Objectives – Options Forgone
Activities – Options Forgone
Built-in-Buts
Double Agents
Shifting Sands

Life may feel like one long battle. In fact it is not. It is a long series of individual battles. One damn thing after another. Therefore it is a very serious mistake to be drawn too deeply into any one battle. We would be very ill-advised to seek to eliminate all dangers. We would have to attempt to catalogue and counter every conceivable danger or risk. That equates to an attempt to match the resources and cunning of the Gods. The immensity of the task would consume all our resources. More important, it would prevent us from fighting the other battles. We would lie like beetles on their backs, legs waving frantically, fielding the incoming risks but making no progress in any direction. What we must do is to prioritise, and then to mount a valiant and thoughtful defence of our activities, our lives, and our assets.

We most certainly will not surrender without a fight. There could be no worse start to a supposedly valiant defence. It would amount to that most craven of capitulations, the ready acceptance of victimhood. If you want to be a loser, that's where to start.

Our strategy must be to offer the Gods as few opportunities as possible to make mischief. The Gods take the same view of a tempting opportunity as most men take of a *femme fatale.* They know they shouldn't. But they know they will. In some ways the Gods are endearingly human.

THE LAW OF UNEXPECTED CONSEQUENCES: THERE ISN'T ONE

The terms 'unexpected consequences' and 'unintended consequences' are often used interchangeably.

Conceptually, unexpected consequences are the more useful. They will always also be unintended. They are failures of foresight, which is a fault which we can address. Addressing the failure of intention would be more of a poser. Therefore we will say no more about unintended consequences.

There is a pecking order of laws. A scientific law is pecker #1. It describes a pattern of nature which has been scientifically observed and identified. It states that, when certain *conditions* are met, invariably there will be a certain *outcome.* Law of Gravity: if an apple is on an apple tree and its stalk breaks, it will experience a gravitational acceleration towards the earth[13].

A general law is pecker #2. Murphy's Laws are general laws which govern human endeavour: if a person or system can go wrong *(conditions),* then sooner or later it will go wrong *(outcome).* They are not scientific laws; the indefinite 'sooner or later' has replaced the definite 'invariably'. But they are general laws because they express certainty: 'will go wrong'.

Is there is a Law of Unexpected Consequences? The statement that 'there always will be unexpected consequences' would be a law, if it were true. But it is not true. The statement that 'there always may be unexpected consequences' is true, but it is a may-be, and a may-be, even a definite may-be, is not a law. It is a possibility. Possibilities are

13 Sooner or later a falling apple will coincide in time and place with the head of a very brilliant man. Sod's Law.

peckers #3, below all laws. There is no law of unexpected consequences to which we are hapless subjects; there is the possibility of unexpected consequences. Laws rule us. Possibilities do not. We can combat them. So come out fighting, expecting the unexpected.

Let us see what we should expect. If we were to be flippant we would say that there are only three things that can go wrong with our plan: the people, the things, and everything else. This may be true, but if we explore both how we go wrong, and how the Gods will help us to go wrong, then we can see how we can make things go right.

THE BATTLEGROUND

The battleground is our castle. We are inside its walls. We are bustling with our chosen activity. The enemy is outside. We wish to pursue our activity all the way to success. The enemy wish our activity to fail; they do not care whether it is through our own ineptitude or by their intervention. The castle has thirteen gates. The gates are spring loaded; unless given continuous attention, they will open.[14]

There are four key areas of risk which determine the success or failure of our activity:

- the risk posed by our choice of objective
- the two risks posed by our choice of activity
- the two risks posed by the enemy's two strike methods, and
- the thirteen risks posed by the vulnerability of our thirteen gates

Let us take them in that order. But first, a quick word on decision-making.

14 Quite why the enemy is hostile we do not know. Perhaps they are unreasonable by nature. We have captured the enemy King's daughter, and our chosen activity is to marry her. She has the only key to his wine-cellar. It is on the same key-ring as the key to her chastity belt. Our objective is to get both. Perhaps that is why the enemy King is jumping up and down. I said that he was unreasonable.

A QUICK WORD ON DECISION-MAKING

We will discuss decision-making in more detail in Part III. For the moment let us agree that the most important decisions are those which – if we get them wrong – absolutely guarantee failure. By that criterion, deciding our objective and deciding our activity are perhaps the most important of the most important decisions.

When making a decision, we forgo the options offered by the alternative decisions. Options forgone are fixed risks which attach to the decision. Not, please note, to its subsequences; they have their own risks which we will come to in later chapters.[15]

Options forgone are analogous to the economists' opportunity costs. Consider them in relation to decisions on resources. All our resources are limited: time, money, energy, manpower and the rest. Resources devoted to the chosen objective and activity are no longer available for other objectives and activities; these alternatives are the options forgone by the decision.

That other options will be forgone is a certainty, a fixed risk. That risk is that it will later transpire that one or more of the options forgone would have been a better choice. This risk must be managed at the time the decision is taken.

CHOOSE THE RIGHT OBJECTIVE

Options Forgone is the one risk attached to our decision in our choice of objective. The risk is that we may choose the wrong objective; the right one is then an option forgone.

How many have metaphorically put a ladder up a tall tree, laboured up to the top, looked around, and oops, wrong tree? By choosing the wrong objective, the wrong tree to climb, we have forgone the option of the right objective, climbing the right tree. Highly amusing to onlookers, less amusing to the climber, and seldom career-enhancing. At least climbing the wrong tree is recoverable.

15 Yes, there is a word 'subsequence'. It means that which *follows* something. Not the same as consequence, which means that which *results* from something. Sunday is a subsequence of Saturday. The Sunday hangover is a consequence of Saturday night.

Cutting the wrong tree down is not. Nor is declaring the wrong war. George Washington had an unfortunate record with cherry trees. Bush minor invaded Iraq to eradicate WMD which didn't exist and to eliminate terrorists who didn't exist until his bungled occupation created them.[16] Clearly in the matter of choosing trees, American Presidents are serial offenders. They should not be trusted with an axe, or a ladder, and certainly not with a dangerous weapon like a bow and arrow.

Deciding our objective may be our most difficult task; it is likely to have to be based more on judgement of the future than on verifiable data: judgement of the difficulties which we can expect, of the difficulties which we cannot foresee, of whether we are likely to overcome them. And most important, of how we will feel at the end of the process, because it is only at the end of the process that we can count our costs and our gains. It is then that we decide whether it was all worth it. There is much to be said for getting the objective right.

Back in our castle, besieged by the enemy, it may later transpire that our chosen objective, the keys, was a bad choice. Maybe one key will not fit. Or both. Bummer either way.

Our objective is inherent in our activity; it is not one the gates in the castle wall. So the enemy will not come pouring in. They will still be outside the gates. Rolling around in laughter as we clamber up the wrong tree.

CHOOSE THE RIGHT ACTIVITY

There are two risks attached to our choice of activity, options forgone and built-in-buts. Built-in-buts we will come to shortly.

16 Terrorists get short shrift in dictatorships. Conversely they thrive in liberal states, and they positively flourish in chaos. The US occupation of Iraq is a classic case of insufficient thinking. The ill-considered plan predictably self-destructed and at the cost of a reported half million lives. Messrs Bush and Blair should be facing the International Court of Justice.

We have just discussed options forgone in our choice of objective. A similar *caveat*[17] applies to our choice of activity. Spend more time at work, have less time for the family, option forgone. Spend money on a new car, you cannot spend it on a holiday, option forgone. Marry girl A, you cannot marry girl B. Not yet, anyway. Spend the day in front of the television eating junk-food, you cannot spend it in the gym, toning your fine physique.[18] By our choice of activity (marrying the girl) to secure our objective (the keys), we forgo the option of obtaining them by another method. Many girls would happily hand them over if you asked nicely and said please. Certainly worth a try.

Options (of alternative activities) forgone are a fixed, overhead risk, inherent in our choice of activity. They are not a gate in our castle wall.

THE ENEMY HAS TWO STRIKE METHODS

Although we have got their King's daughter, we are not certain of the identity of the enemy at the gates. Let's plan for the worst, let's assume it is the Gods outside the castle walls.

They are formidable adversaries. They have an unerring ability to zero in on our deficiencies. They particularly relish punishing any manifestation of hubris. Their first strategy is to encourage us to self-destruct. Their second is to offer a helping hand.

Strike Method 1: Divine Judo

When our endeavours to defend ourselves are ill-considered and easily demolished with little force, the Gods employ their first strategy, divine judo. They lead us up the garden path to self-destruction.

Their first judo technique is to do absolutely nothing to interfere with our activity, in order to nourish our complacency. If we are truly foolish, we will allow complacency to grow into hubris. This will displace our sense of self-preservation, and we will go on gathering momentum until we trip over our own feet.

17 Caveat – watch it.

18 Option forgone, not built-in-butt.

Should we not fall for their opening gambit and trip ourselves up, they offer a helping hand. They lie in wait with their second technique, their grand repertoire of unfortunate coincidences. As soon as we offer them an opportunity, we are undone. An unexpected coincidence will catch us off balance. Whoops, down we go. Sod's Law has struck.

If we succumb to either of these techniques, we have only ourselves to blame. To ignore the dangers of hubris or complacency or nasty coincidence before embarking on any endeavour is to embark on self-destruction. Technique one encourages us to fall over ourselves, no help needed. Technique two assists our self-destruction with a well-placed coincidence. The Gods are black belts. We are losers. Respect for the Gods.

Strike Method 2: Divine Sumo

If we are well-prepared we will not offer so simple a target. We will have constructed a careful system of defences. The Gods are unperturbed. They turn themselves into heavies. They adopt their second strategy, divine sumo. Once more they hang back and await their opportunity. When they see it, this time they hit big. Wallop. Lightning strike. They send a thunderbolt through our whole system of defences in a flash, allowing us no opportunity to organise a blocking response. Their attack is fast, heavyweight, and unstoppable. It is at an unexpected time and from an unexpected direction. Our system defences fall like a row of dominoes. Bang. Down we go. We plead in our defence unexpected circumstances. Pathetic. When you are up against the Gods the unexpected is always to be expected. We are losers again. The Gods are sumo masters. Big Gods. Big respect.

GATES ONE TO FOUR: SELF-INFLICTED INJURIES

The first four gates in our castle wall are labelled self-inflicted injuries. They are the gates with the most powerful self-opening springs. They are our biggest source of risks. We will do well to keep an extra-careful eye on them.

- Gate one is ignorance of the rules of battle

- Gate two is breach of the rules of battle
- Gate three is failure to understand the forces arraigned against us
- Gate four is failure to deal with the forces arraigned against us

Self-inflicted injuries proffer to the Gods opportunity for mischief; they provide them with their favourite avenue of attack. Most self-inflicted injuries are acts of predictable stupidity. The Gods know that they can rely upon us for lots of them. They are the most common cause of our undoing. To avoid self-inflicted injuries, we must keep gates one to four firmly closed.

To keep gate one closed, we must understand all the rules of battle.

To keep gate two closed, we must adhere to the rules of battle. This means that we must conduct our affairs with the greatest possible competence. For a start we must keep gates three and four firmly shut. We know that the Gods have a predilection for mischief. If our failure to follow the rules offers them opportunity, they will be sorely tempted.

Failure by reason of incompetence can be avoided by diligence. Failure by reason of complacency the Gods will see as an open invitation. They are very likely to accept it. Failure by reason of hubris is too terrible to contemplate. It is worse than offering the Gods an opportunity, it amounts to insulting them. 'Hee, Hee, Hee, you can't touch me', are not words to use to a thunderbolt launch-team.

We know that, even with our best efforts, the Gods *may* choose to zap us. But without our best efforts, they undoubtedly *will* choose to zap us.

GATES FIVE AND SIX: MURPHY'S LAWS

We understand Murphy's First and Second Laws. Careful observation of the risks which they present should enable us to keep these gates closed.

GATE SEVEN: INSUFFICIENT THINKING

We will discuss *insufficient thinking* in detail in Chapter 13. For the moment we will define it as the failure to identify all factors crucial to our situation. Keep an eye on gate seven.

GATE EIGHT: SOD'S LAW

We understand Sod's Law. We will watch gate eight like a hawk, ready to swoop on to any nasty coincidences trying their luck.

GATES NINE AND TEN: OBJECTIVES AND ACTIVITIES FORGONE

Gates nine and ten are not gates which may let the enemy in. They are gates at which a steady stream of informants calls. We man these gates with intelligence officers. Their task is to listen to what the informants have to say about our chosen objective and our chosen activity. Informants will mostly have misleading intelligence which, if acted upon, will disrupt our operation. Very occasionally their intelligence – that we are pursuing the wrong objective or the wrong activity – will be spot on. Then the intelligence officers will have the career-busting task of informing the boss that he has selected the wrong tree or the wrong ladder. To add to the fun, the enemy may send informants with disinformation.

GATE ELEVEN: BUILT-IN-BUTS

Built-in-buts are our second risk. They are the disadvantages which attach to any given course of action. As previously stated, cyclists get cold; motorists get fleeced by petrol prices. Our success is determined by the outcome of the tussle between our management of the disadvantages and the Gods' endeavours to exploit them. Built-in-buts are a gate in our castle wall which we must keep closed.

Our chosen course of action, marriage, has a number of built-in-buts. If you would like a list, ask your spouse. On the other hand – if I may make a brief excursion into marriage guidance – don't. One secret of a happy marriage is to keep off tricky subjects.

Because built-in-buts are certainties, we can argue for the existence of a law of built-in-buts. It governs, if not all nature, then certainly all life. The disadvantages of being a tiger, predator king of the jungle? You can't live on leaves, so if there is no prey around, you starve. The disadvantages of being stinking rich? Tricky, I can't draw on experience. Offers of a peerage for a suitable consideration? 'That will do nicely; thank you, my Lord.' The disadvantages of being tall, dark and a handsome? No experience again. Women don't appreciate your fine mind? The disadvantages of being a stunningly beautiful woman? Er, research subjects needed.

Medicines provide clear illustration of built-in-buts. Take aspirin and Viagra®.[19] The instructions for use of any medicine will include a list of possible side-effects. Aspirin may irritate your stomach; it may endanger small children. It may also thin your blood. And now things get interesting. Just as advantages have their obverse side: disadvantages, disadvantages also have their obverse side: advantages. The disadvantage of thinned blood turns out to be an advantage to those with weak hearts; it is easier to pump around. Viagra was originally intended by its developers to be a heart treatment drug. One of its built-in-buts or disadvantages turned out to be regarded by many men as a very considerable advantage. The danger that its users may then give themselves heart attacks in the course of their exertions completes the circle. Unintended consequence indeed, but perhaps to be expected.

This dichotomy, that every action has its downside, its built-in-buts, is very interesting. We will look more closely at it when we discuss strategy. To get the flavour, consider the textbook military dilemma. The enemy is in rapid retreat. How far do we pursue him? Hold fast and reorganise, but perhaps miss the opportunity to destroy him? Or pursue him at full force, but risk a successful attack on our stretched and unguarded lines of supply? Or leave guard forces along our supply-lines, but risk heavy counter-attack on our depleted vanguard? The law of built-in-buts is nature's law of negative feedback. It is one of the universal laws of

19 Example, not prescription.

nature. Everything carries the seeds of its own destruction. More on universal laws and on strategy in Chapter 13, Thinking – Join Up Now.

CASE STUDY 2.1 BUILT-IN-BUTS: LOOK AROUND YOU

Consider your own life; you will see built-in-buts all around you. They are inescapable. I live in an old house, oak beams, inglenook fireplaces, everything that makes life wonderful. And yes, I regularly drink a libation to the Immortal Gods humbly to acknowledge their benevolence. No complacency here.

Built-in-buts come with a period house: they have small windows and are dark inside. (House with big windows: option forgone.) And whilst I will not admit to being depressed by anything, I will admit that I find sunlight very cheering. For that reason I decided to write this book sitting in a bright and sunny summerhouse. I bought myself a laptop with a nice big screen.

Built-in-buts come with the summerhouse: the brightness of the background light and the multitude of reflections make it very difficult to read the nice big screen. So I bought a shade device to fit on the laptop, £20. Useless. So I bought an old fashioned folding screen. Ex-factory cost in China, probably £20. Cost in UK, nearly £200. Some help but in a confined space a pain to adjust. Solution: place laptop and would-be author's head in a large cardboard box.

Built-in-buts come with the chosen solution: quiet sniggers of my nice neighbours, loud derision of my dear children, demands of more maintenance from ex-wife on grounds that long-alleged mental state now proven beyond reasonable doubt.

But do the built-in-buts matter in this case? No, the operation has not been disrupted, no one has been hurt. *De Minimis Non Curat Lex* – don't sweat the small stuff. My feelings injured by the derision? Irrelevant. If you can't take a joke, you are on the wrong planet. Did I fail to expect the unexpected? Yes, but it was not of great import, and how long should you spend faffing about, researching trivia? Not long. The objective of planning is not the perfect plan. It is to inform action.

The risk arising from the built-in-buts of our chosen activity is that, if not managed successfully, they will prevent us from attaining our objective. Our first step is to try to identify all important built-in-buts. This will remove them from the list of unexpected consequences to the list of expected consequences. We can then plan how to deal with them.

GATE TWELVE: DOUBLE AGENTS

The twelfth gate in our castle wall is the one labelled double agents. It is the one on which we think we can rely. Therein lies its danger.

Double agents are our own risk management measures. Unfortunately they always provide new sources of risk.

The observation that any activity carries its own risks applies equally well to the activity of risk management. The safer you feel, the more carefree you become, the more you push your luck. Or, put another way, the more risks you may be prepared to take. This is called risk homeostasis. Would you be willing to stand on one leg and close your eyes in the middle of your lawn on a nice day? Probably yes. Would you be willing to do the same at the edge of a precipice on a windy day? I hope not. In such conditions you would be more careful.

Every one knows that motor cyclists are safer wearing helmets. Certainly they are better off with a helmet, if they fall off. But are they better on with a helmet? What if wearing a helmet makes them feel safer, encourages them to ride faster, and makes them more likely to have an accident? There are some very careful studies which suggest that this is the case, that the helmet acts as a double agent . It is a neat paradox; better off the bike with a helmet, better on the bike without a helmet. For the same reasons, doctors believe that if rugby players affected the same massive protective gear as American footballers, they would suffer not less but more and worse injury. They would also look silly.

All risk reduction measures are double agents. They bring with them three additional risks.

- The first additional risk is that the additional safety measure is itself a new source of error.
- The second additional risk is that too much, even total, reliance may be placed on a safety measure. The classic case is the Maginot Line, the impregnable front gate fortifications upon which France relied to keep Hitler out. Unfortunately the French did not think to lock the back-gate, Belgium. Reliance on a safety measure leads to the decrease in vigilance (complacency), the removal of back-up systems (complacency and cost cutting), and when the infallible safety measure is by-passed or fails (sooner or later it will – see Chapter 3), bang. Total reliance on any safety measure is a self-inflicted injury. Don't do it.
- The third additional risk is that the additional safety measure adds complexity. This complexity adds to the operator's difficulty in understanding the system. And he is the last resort, the chap who will have to sort it out when all else has failed.

These three risks accompany any safety measure. They are the double agents who may work for the other side. Expect them, or expect the Gods. They will come sailing through the open gate.

CASE STUDY 2.2 DOUBLE AGENTS: SECOND PAIR OF EYES

Adding another pair of eyes to an important activity would appear to obtain a simple and immediate enhancement of safety. Unfortunately it has often had the opposite effect; the individual vigilance of each participant has reduced because each has been relying on the expertise of the other. For it to have a positive and beneficial effect, the roles of the two participants must be carefully defined, learned and followed.

Most modern airliners are operated by a crew of two pilots. One of the bedrocks of safe aircraft operation is the continuous checking by each pilot of the other's intentions and actions. It is called mutual monitoring. It is exactly what happens on airlines which insist on the highest professional standards. Mine did. Many don't. Avoid them. Extract from an RAF Board of Inquiry report into a fatal accident: ' … as it is unlikely that [the first officer] would have had reason to doubt the actions of his experienced captain, … ' Absolute rubbish. That statement, and the fact that it was not sent back for reconsideration by the air officers reviewing the report, suggests that the RAF at its highest levels had not grasped the essentials of mutual monitoring. For this accident, the RAF blamed its two dead pilots. Monstrous injustice.

Extract from an investigation into a medical accident: 'I was a SHO [junior doctor] and did what I was told to do by the Registrar [senior doctor]. … Dr X was employed as a Registrar by [hospital Y] which is a centre for excellence and I did not intend to challenge him' He didn't. The patient died. For this accident, Dr X was jailed. Monstrous injustice.

You will see why these verdicts are wrong in Chapter 7, where both accidents are reviewed.

GATE THIRTEEN: SHIFTING SANDS

Gate thirteen is built on shifting sands, hence its name. Sometimes the sands move and it springs open. Sometimes it is inadvertently opened by the people around it, trying to make themselves more comfortable.

Shifting sands are the risks occasioned by changes in the environment of our operation. The world is in a natural, perpetual state of change. Also and worse, it changes in response to any activity. Times change, customs change, regulations change, rivals emerge; the social environment closes around plans and activities as a predator closes around its food, seeking to extract the bits it likes and to reject the bits it does not like. People, groups and organisations always utilise any opportunity to maximise their own advantage. Unfortunately any plan of activity is essentially static. Therefore there are ever-present risks posed to the plan by the dynamism of the world in which it is intended to work. These shifts in the sands are the main causes of so-called unexpected consequences. We should expect them.

The more complex the plan, the greater the number of elements in its environment. Each element brings with it its own risks, including that of shifting sands. A simple domestic plan to have a quiet cup of tea in thirty minutes' time by oneself is not at great risk of shifting sands. Certainly the telephone might ring, mother-in-law might drop in, we might be out of tea, or the kettle might be on the blink, etc. All these possibilities are manageable.

The more complex the plan, the more the possibilities of sands that might shift, and the more the challenge to our management skills. If our plan were to have a cup of tea not alone, but with four friends, each one of the four would introduce further risks to the plan.

Additionally, the longer the duration of the plan and its associated activity, the longer the period in which the sands can shift. Plan the cup of tea not for thirty minutes' time but for thirty days' time and the odds on success are markedly diminished. If instead of a quiet cup of tea with four friends, we intend a dinner-party of ten courses for twenty guests, followed by a performance of *Aida* in the garden (as one does), each additional element will introduce an additional source of shifting sands. *Aida* will of course introduce shifting elephants. And each additional elephant will introduce, etc.

Shifting sands are the main cause of unexpected consequences. How do we manage them? Prayer does not constitute management. Panic is not the right answer. The essential is a most careful and sufficient analysis of the likely responses to the plan, that is of which sands will shift, and how they will shift.

CASE STUDY 2.3 SHIFTING SANDS: AT A HOSPITAL NEAR YOU

Now consider a plan set not on a domestic stage with a handful of actors and a short time-span, but set on the national stage with a multitude of actors and a longer time-span. More people, more interest groups, more things, more systems, and more time for the sands of all of them to shift. The government decides that more money is the answer to the all-pervading ghastliness of the National Health so-called Service. In goes £10 billion, but with no attempt to forecast and control the shift in the sands which will be caused by this plan. One interest group, the doctors, shifts fast and collars the lot. Result: doctors laughing, plan up spout, NHS unimproved.

A few months on, the top doctors, the consultants, are being asked to cut back on their hours (they are paid by the hour) to save money. Fewer consultants' hours equals fewer patients treated; ghastly service made more ghastly by the measures intended to improve it. Unexpected? Absolutely not. Totally predictable. Unexpected only to those who could not find beer in a brewery.[20]

And the next year the House of Commons Public Accounts Committee reported that government-decided changes in another health area, out-of-hours visiting, had served to benefit only the doctors.[21]

The first step in the analysis of shifting sands is to list all the factors in the environment of the plan and its activity. For example, let us plan for a small picnic. What can shift? Weather, venue, food and drink, transport, people. Weather: possible responses – marquee, or postpone until

20 Unfortunately such a one was in charge. Ironically it was the government's apology for an attack dog, the Scots terrier John Reid.

21 Doctors seem to know what they are doing. Should they take over the government from the politicians? Answer: No. They are better at looking after themselves than after us. Just like the politicians.

weather favourable. Venue – alternative venue. Food, drink and transport – logistics, not difficult. People – again not difficult.

The wider-ranging, the more complex the undertaking, the more difficult it will be to identify and list all the factors. It will be necessary first to identify their broad categories: environmental, economic, political, social, commercial, technical, individual, etc. Within these categories we will then identify the major factors that will change either over time or in response to our plan. Of particular importance are the interest groups, those with an axe to grind. Some of them may see our plan as inimical to their interests and will actively oppose it. There may well be others, supposedly working for us, whose real interest does not lie in our agenda, the successful outcome of our plan; it lies in their own agenda and the attainment of their own objectives. Some will be obvious: identified interest groups, known turf wars. Some will be undeclared: senior directors plotting their next corporate theft, managers scheming a buy-out, senior RAF officers desperate for their knighthood, *et al.* Some of the opposition to our endeavours will be open. Some will be clandestine and stealthy.

The simple passage of time will also bring changes to the environment. Hostile people may become friendly. Friendly people may become hostile. Old problems may be solved. New problems may emerge. New people and new products may appear on the scene. Fashions will change. Sands will undoubtedly shift. Predicting the details of how they will shift is an entirely different matter.

Getting to grips with shifting sands is a tricky operation. Grip them and they shift again. Therefore, remember that the sands are always shifting. Another gate to watch.

CASE STUDY 2.4 SHIFTING SANDS: FRIENDLY FIRE

Scott Snook's gripping book, *Friendly Fire*, illustrates the gradual divergence of practice from plan. Scott Snook is a Lieutenant-Colonel in the US Army, and a Professor at the

US Military Academy. A combat officer, he had his own interesting experience of being shot up and wounded by the US Air Force. His book is an accomplished analysis of the incident in which two US Air Force F–15 fighters shot down two US Army Black Hawk helicopters, killing all 26 crew and UN peacekeepers on board (14 April 1994).

The helicopters were in a zone which regulations had made a no-fly zone for friendly aircraft. So the F–15 pilots saw them as hostile Iraqi Russian-built Hinds, and shot them down. Simple.

No, not simple. The no-fly zone had been routinely ignored for some time as a matter of local need and practice, but no-one had informed the F–15 pilots of this.

Snook identifies this mismatch between dated regulations and current practice as a major component in the causation of this accident. He gives the name 'practical drift' to this tendency of current practice to depart from original plans.

In our terminology, practical drift is one manifestation of shifting sands. The US Air Force's failure to manage them was one cause of this tragedy.

There was of course more to it than that. A US AWACS (surveillance and control) aircraft had an overview of the scene, and was in contact with F–15s, but failed to intervene. The F–15 pilots failed to note the distinctive difference between the colour and paint scheme of US Black Hawks and the supposed Hinds. And more. Actually, rather too much more.

We will return to *Friendly Fire* in our discussion of accident investigation (Chapter 7).

The prediction of exactly how sands will shift can be extremely difficult. The more complex the activity and the longer the time-scale, the greater the difficulty. In human affairs and at national and international level it is almost certainly impossible. Hence the efforts which organisations make to take control of their environment. The USA seeks to control the world – further disastrous blunders to come. Commercial organisations seek to dominate markets. The purchasing power of large organisations enables them to undercut smaller competitors and ultimately to drive them out of business. The more powerful they are, the more powerful they become. Tesco is an obvious example.

Shifting sands underlie the civil servants' habitual caution. If they delay, many of the sands will have had time to shift, and there is more predictability of outcome. Unfortunately, delay has a big built-in-but. Opportunities are missed. Politicians may delay in addressing a small problem because it is one small problem amongst bigger problems. The problem may fade away, as they had hoped. Or it may grow, and then they have will missed the opportunity to kill it before it got too big to be killed. The politicians' defence is that the public will not support firm measures against small problems, so they have to wait until the problem is big enough for the public to see. It is then too big for the politicians to solve. Terrorism is a good example. Managing the problems posed by shifting sands is how supposed statesmen and politicians should earn their money. They don't earn it. But they still take it.[22] Join the Campaign for Real Statesmen. The Riddle of the Shifting Sands is too much for the present lot.

RULES OF BATTLE

1. Know these rules
2. Observe these rules
3. Understand the forces arraigned against you
4. Deal with the forces arraigned against you
5. Self-inflicted injuries are your biggest risk. You incur them if you fail to observe these rules. Serves you right
6. Complacency and hubris are certain routes to self-inflicted injury (see Rule 5). They must be banished from your mind, absolutely and permanently

22 And then make a fortune from their self-congratulatory memoirs.

7. Insufficient thinking is the third route to self-inflicted injury
8. Sufficiency of thinking prevents unexpectedness of consequences
9. Murphy is not all there. But he is always there
10. Sod is ever-ready. He has a wide range of nasty coincidences for any occasion
11. When you take a decision, alternative options are forgone
12. Try to choose the right objective
13. Try to choose the right activity as your means to your objective
14. Built-in-buts accompany any activity. They are its inherent disadvantages
15. Double agents are the risks posed by your own safety measures. Given half a chance, they will work against you
16. Shifting sands are risks posed by changes in the environment of your system of operation. The environment changes naturally, and in response to your activities
17. The Gods encourage complacency. They lead you up the garden path to destruction. Don't go
18. The Gods attack your systems of operation with a heavy, lightning-fast strike. It goes through your defences like a thunderbolt. No surprise. It is a thunderbolt
19. You can manage these risks. If you wish to win the battle, you must manage them

So there we are, Ladies and Gentlemen. Outside the castle walls, the Gods, mischief-makers *par excellence*, armed with the Murphy's ubiquitous Laws and with Sod's Law of nasty surprises. Inside the castle, us, determined to win, cognisant of the rules and of the enemy, gates firmly closed, expecting the unexpected and with a rapid response of fancy footwork in reserve.

Let's hope the keys fit.

CHAPTER 3

A TOE IN THE WATERS OF RISK MANAGEMENT

QUICK RECAP

Our concept of risk is simple: life is one long tussle with the Gods; the rules of engagement are vexatious; we are handicapped by our human fallibility at all times; the Gods are empowered to make mischief at any time.

Our objective is to formulate an action plan for the management of the risks of our lives. At first sight it may appear that we are faced with a huge and complex task. But have we not already determined to be valiant and thoughtful in our endeavours? Yes, we have. So are we daunted by the insurmountable and the infinite? No, we are not.

- First, we get to grips with the complexity of the situation by thinking about it sufficiently, in other words both widely and deeply
- Second, we take great care with the formulation of our action plan; we particularly strive to exclude self-inflicted injuries
- Third, we accept that we will reach a limit to our understanding. At that point we will no longer be able to predict with reasonable accuracy what we may face
- Four, we do not agonise about the unknowable. That would serve only to paralyse us
- Five and most important, we do not go to the opposite extreme. Well aware that the road to hell is paved with hubris,

we certainly do not press on, blindly complacent that nothing can go wrong

- Six, when we reach the limit to our understanding of the situation, we move from the construction of action plans to the construction of contingency plans. They will contain our initial response to nasties, both known and unknown. Obviously, we cannot formulate a comprehensive response to unknowns. The function of contingency plans is to limit damage and to buy us time in which to devise a detailed response

Murphy's Laws tell us that if our activities can go wrong, sooner or later they will. We know that if we are careless we will trip over our own feet. If we are complacent and offer the Gods an easy opportunity, they will upend us with divine judo. If we prepare a system of defences, their attack will be by divine sumo, a thunderbolt of coincidences, aka Sod's Law. Knowing what we face is an excellent start to the planning of our endeavours.

We know that risk attaches to everything, to our well-being, to our activities and our assets. Before we dive into the management of all these different risks, we must first agree on our terminology. The words risk, asset, hazard and loss are the tools which we will use to dissect[23] the task of risk management. A simple model of this task would also be helpful.

Asset risk is less complicated than activity risk; we will use it to sort out the terminology and to provide us with our first model.

LOSS

Do all the schools of thought agree on the meaning of risk? No, they don't. On the meaning of hazard? No again. Do they agree on anything? Never mind, we will agree our own meanings. Let us start with loss.

Loss is a simple now-you-see-it-now-you-don't. In material or tangible form it could be the loss of the £25 which you put on the 1430 hr

23 Actually vivisect, it is a live problem.

at Kempton Park, of your new car which you unwisely loaned to your teenage daughter, or of any material asset which you care to name.[24] In less material form, it could be the loss of privileges – your golf club membership; of rights – your parking space; of prospects – of your promotion, of your job (let's hope not), or of some other semi-tangible. In human form it could be the loss of any source of satisfaction – of good reputation, of well-being, of happiness, of good health, or of life itself. Some assets stretch across more than one category. The loss of employment may involve tangible, less tangible, and human (sickness, depression, stress) losses. Each of these putative losses relates to (the enjoyment of the benefits of) an asset in one form or another, whether the enjoyment is present, or prospective, or both. Obviously loss may be total or it may be partial. After Captain Clot's heavy landing the aircraft was a total loss. Captain Wheels-Up's landing damaged the fuselage but not the wings, nor the wheels because they were safely tucked up in the wings.

Two concepts underlie our understanding of loss.

The first concept is of an *asset*, loosely defined. Remember that our interest is in avoiding misfortune, therefore by our definition an asset is a Good Thing. It is beneficial. More correctly stated, our first concept is *the existence of the benefits deriving from an asset.* However we do not wish to disappear into metaphysical niceties. So if we understand that existence and benefit are inherent in the term 'asset', we need not keep referring to them. Nor will we involve ourselves in the question of ownership. It is a complication which is not central to our broad discussion. Suffice it to say, if you have a lovely view from your house, you will probably regard it as an asset. But unless you are stinking rich, you probably will not own it. Nonetheless you will derive benefit from its existence and, if some delightful developer builds a block of flats on it, you will suffer a loss. Sadly the developer will not. He will go off and buy himself an unspoilt view which he will then defend tooth, nail and backhander. Bring back the gallows.

24 Why does it cost $millions to divorce a Jewish-American Princess?
Because it is worth it.

The second concept is the removal of the benefits of the asset. And there we have our understanding of loss. *A loss is the removal of the benefits of an asset.*

RISK

In loose usage, the word risk is often used as synonym for probability (of misfortune or loss). There is an 80% risk (probability) that the builder will not turn up next week. My daughter requested a loan of £1000 for a holiday that she insisted she needed (needed?). There is a high risk (probability) that she will view the loan as a gift. This usage of 'risk' is unhelpful. A word is a unit of communication. Ideally it should convey its meaning simply, clearly and unambiguously. There is absolutely nothing to be gained by replacing the clear, single meaning of the excellent word 'probability' with the confusing mish-mash of meanings of the woolly word 'risk'. That is, there is nothing to be gained by honest men.[25]

For risk management professionals the word 'risk' has a specific meaning. Their 'risk' is a man-made concept. It is the product of the amount and probability of the possible loss, in other words, the stake and the odds on losing it. To be precise, their definition of risk is the magnitude of the possible loss multiplied by the probability of the occurrence of that loss.

Risk = Magnitude (of Possible Loss) × Probability (of Occurrence of Loss)

So if you bet 10p on the toss of a coin, the magnitude of the possible loss is 10p, the probability of its occurrence is 50%, and your risk is 5p. If you increase the bet and the magnitude to £10,000, then you up your risk to £5000. The concept underlying this definition is useful. It provides a measure for three essential tasks of life: the assessment of risks, the comparison of risks and the act of choosing between them.

25 Ambiguous words do have their uses, good and bad. When they provide fun for wordsmiths and their readers, no one will object. When they provide a means of deceit to the corrupt, politicians, spin-doctors, and greedy money-grubbers, we all object, and quite right too.

At the extremes of magnitude of loss this calculation is not informative. The loss of 10p, even if highly probable, is of no consequence. At the other extreme, complete financial wipe-out, even if supposedly highly unlikely, is to most of us a disaster to be avoided at all costs. Not so long ago, many Lloyd's names found this out the hard way. It matters not whether they were victims of misfortune or of deceit; their contract exposed their total assets to both. At the extremes, any risk calculation is likely to be spurious, if not plain dangerous. Its appearance of accuracy will serve only to obfuscate a simple commonsense decision. Furthermore, the quantification of risk is itself a risky business, as we see later in this chapter. The first question is the magnitude of the possible loss. If it lies at either extreme, trivial or catastrophic, do not reach either for your calculator or for risk consultants; the decision is already clear. Only if the magnitude of the loss does not lie at either extreme should we proceed to the second question, that of the risk.

HAZARD

There are various definitions of hazard to be found in the literature. The question to be asked is not: which one is right? That merely exchanges a dispute on meaning into a dispute on the means of resolving the dispute. Let us duck this issue by devising our own definition. Our criterion is not the abstract correctness of our definition but its practical utility.

We are seeking to construct a mind model of our real-life task, the management of the risks of life. We have some building blocks in place, the concepts of an asset, of the loss of its benefits, and of the associated risk – magnitude and probability of loss. What we need now is a concept of the source of loss. For the moment we are carefully avoiding the word cause, because the nature of the cause of a loss or accident is a dispute into which we will wade with enthusiasm in our next chapter.

So, the concept of a source of loss. The simplest source of loss is pure chance, the toss of a coin, but only if it is absolutely sure that the coin is not weighted and the tosser is not, er, a tosser. Games purportedly of chance but with well-stacked odds are a time-honoured means of fleecing the innocent. What are the possible reasons for the loss of a simple asset? Consider, for example, your wrist-watch. The mechanism may fail, perhaps as a result of wear and tear, or of poor design; the watch may suffer accidental damage – you may drop it, or immerse it during washing-up duties; it may be stolen by an undesirable. Behind each possible reason for loss are one or more underlying sources. The source of wear and tear may be lengthy usage, or poor maintenance, or poor design. The source of accidental damage is likely to be human behaviour, perhaps error; it may be aided and abetted by chance. The source of theft is a thief. We are not seeking to compile a comprehensive list of each source of loss. Nor are we seeking to analyse the sources further, to identify sub-sources. We seek only to illustrate that there will be at least one source underlying each reason for a loss. In other words there can be no loss without a source of loss. Source of loss is therefore a concept both necessary and useful to our model of the management of risk. In our proposed terminology *a source of loss is a hazard.*

We must distinguish between a *hazard* and a *potential hazard.* An earthquake which is occurring is a hazard. An earthquake which may occur is a potential hazard.

Our model of asset risk so far has three structural components: an *asset,* a *loss* of asset benefits, and a *hazard* or source of loss. It has one associated interpretive measure, *risk.* One structural component is missing. What is it? Hint, if there is a loss, there must be …

Quite right, a loser. The unfortunate who called heads when it was tails; the owner of the blighted watch; the *ingénu*[26] who expected the repayment by his daughter of his loan to her. But we are not going to use the term loser; that word is too useful as a description of the chronically incompetent, and of those miserable creatures whose sole pleasure in life is the expression of anguish. Their perpetual air of martyrdom tells the world that they are suffering terribly but being oh so brave. Light the fire, somebody. That's not us. We are winners, hard-wired with PMA – Positive Mental Attitude. We are not losers;

26 Mug. But let's not be personal.

we are stakeholders. The concept of a stakeholder is wider and more fundamental than that of loser. In all but the simplest losses there will be more than one, often many, stakeholders. The loser is amongst them. Some stakeholders may stand to gain. Following a loss, the interests of the different stakeholders will almost certainly conflict.

CASE STUDY 3.1
TRY NOT TO MEET MRS JONES

The Stakeholders in a Hypothetical Motor Accident

Mrs Jones is taking her daughter to school; she is late
She is driving too fast down a single track lane
Miss Smith comes the other way on her horse
Mrs Jones cannot stop
Miss Smith and her horse are injured
Mrs Jones' daughter is shocked
Miss Smith's parents are incandescent
Mr Jones is incandescent; this is not Mrs Jones' first accident
A firm of professional liars has been retained to assist her
Two insurers are involved
The police are involved
Lawyers are involved
Twelve stakeholders already, and still counting

FIRST MODEL OF ASSET RISK MANAGEMENT

Our first model of asset risk management is now complete. It has four components:

- a stakeholder
- an asset
- a hazard
- a loss

It also has one tool, a measure which assists our understanding of the model:

- risk, the magnitude of the possible loss multiplied by the probability of the occurrence of that loss

Remember that the interests of the stakeholders may be different, in which case we will need a separate model for each stakeholder.

MODEL 3.1 FIRST MODEL OF ASSET RISK MANAGEMENT

STAKEHOLDER

Hazard **Asset** **Loss**

This is a very basic model. It does not tell us anything complicated. It is nonetheless useful in that it identifies the four major components in the system and five spaces between them. Each component and some of the spaces offer a risk management opportunity, an opportunity to avoid or minimise the loss. If you would like to try to work them out for yourself, cover up the next paragraph and pause here. Otherwise read on.

Model 3.1 assists us to identify our risk management opportunities:

- The Stakeholder could relinquish or reduce his/her stake
- The Hazard could be removed or reduced
- The Asset could be removed completely or in part
- The Loss could be reduced, e.g., by duplication or back-up of important features
- In the Stakeholder–Loss space, insurance could be inserted
- In the Hazard–Asset space, asset protection measures could be inserted

SECOND MODEL OF ASSET RISK MANAGEMENT

Our Second Model is still simple, but it provides some further and useful insights into our task of risk management.

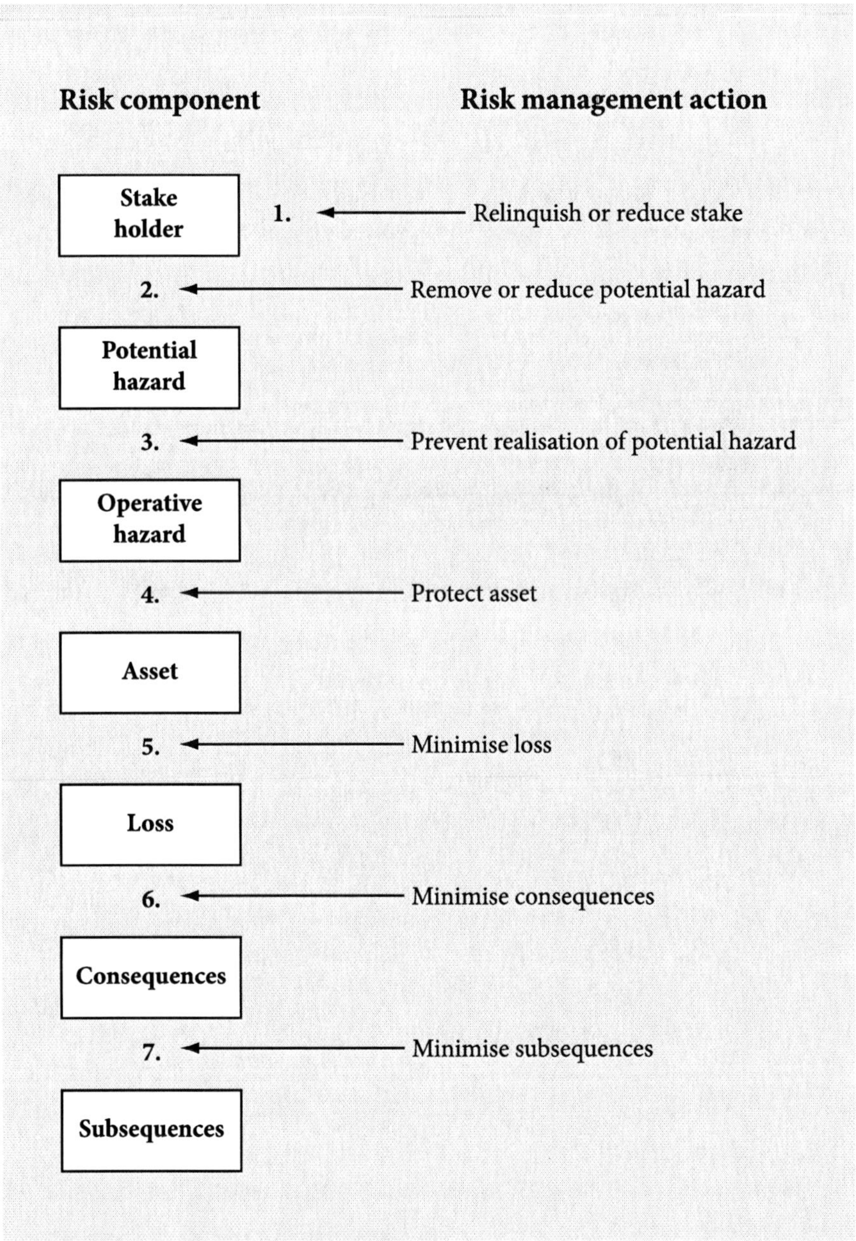

THE LESSONS OF THE SECOND MODEL OF ASSET RISK MANAGEMENT

Each space between the individual modules of the model offers a risk management opportunity. Endeavours can be made to disrupt the process of system failure by severing or restricting one or more of the links in the sequence. Similarly, defensive measures can be taken to reduce the loss, the consequences and the subsequences.

One *stakeholder* is the potential loser. Risk management should start by assessing his position. If his stake can be eliminated then – from his viewpoint – the risk management task will have been completed; if his stake can be reduced the subsequent risk management task will be substantially altered.

For example, if the risk which is being addressed is the impact of an earthquake, the stakeholder's theoretical first options are to relinquish his stake by moving out of the earthquake zone, or to reduce it by moving farther away from the epicentre. As before, a separate model may be needed for each stakeholder.

A *potential hazard*, if realised, becomes an *operative hazard*; there may be opportunity to prevent this. Thus the possibility of an earthquake is a potential hazard; the occurrence of an earthquake is an operative hazard. The prevention of natural calamities may not be practicable, but most other potential hazards can be addressed successfully. An obstruction on a motorway is a potential hazard which can be eliminated by its removal, or which can be reduced by warnings or by cones.

If the potential hazard cannot be prevented from becoming an operative hazard, the attention of risk management moves to the construction of defences around the *asset*. Buildings can be reinforced in earthquake zones; defences put in place around terrorist targets; cars can be strengthened to protect passengers.

Minimisation of the *loss* is the next risk management opportunity. Insurance cover can be purchased. Valuables can be moved from the house in the earthquake zone. The number of key persons on one aircraft can be strictly limited so that a single fatal accident does not result in the total loss of specialist expertise. Aircraft can be designed so that the loss of one minor system does not cause the loss of the whole aircraft.

A loss has immediate *consequences.* They reflect the situation at the time of the loss and are then unavoidable. Prior to the occurrence of the loss, consequences are susceptible to risk management. An earthquake will have grave consequences to a stakeholder if his house is not insured and if it contains the only copy of his *chef-d'oeuvre.* They will be less grave if he has insured his house and deposited a back-up copy of his masterwork in a safe place.

A consequence to a commercial organisation of the loss of a key asset may be its failure. The same loss may be of little consequence if its risk has been well-managed, for instance if a sound business continuity plan is in place. The consequences of an aircraft engine failure may be catastrophic if the pilot is not competent. Such consequences can be avoided by simple risk management before the event, by ensuring that the pilot has been properly trained and can operate to a professional standard.

Subsequences (in this context) are the overall impact on the stakeholder over the long term. Whereas consequences and the earlier components of the risk can be managed only *before* the loss, subsequences can and should be managed *both before and after* the loss.

The importance of subsequences lies along a scale from nil to overwhelming. For those who lose their lives, there are no subsequences. For those who suffer bereavement, subsequences may be insignificant in comparison to their loss. However, to most stakeholders the subsequences may constitute the single most important component of risk management.

The subsequences determine the survival or otherwise of the organisations and individuals involved. It is then that recovery action is taken, events are investigated, responsibility is apportioned, reputations and jobs are lost. It is then that history is written. Or rewritten. Tony Blair was accused of responsibility for the losses and consequences of his invasion of Iraq (2003). He owed his lengthy survival in office primarily to his, or more likely, his press secretary's, clever management of the subsequences. We may note in passing the considerable skills of contemporary politicians in managing the subsequences of the losses which they occasion us. We may regret that their skills do not extend to managing anything else.

MAPS AND CHECKLISTS

In broad terms, risk management methods come in two forms, mapping and checklists. Mapping draws a diagram of the components of the system. Checklists give step-by-step instructions. Maps assist visualisation. Checklists are logical pathways. The nature of the situation will determine which is the more appropriate. The nature of your mind will determine your personal preference. Both methods can be helpful, but only if well thought out. If superficial they are, like half-baked quantification (which we will come to in a moment), plain dangerous; by giving a false illusion of certainty and understanding, they are a first step on the road to ill-founded complacency.

It is important to remember that a model can never be more than an aid to thinking. It should never ever be used as a substitute for thinking. If any owner of a proprietary model seeks to persuade you otherwise, direct him to the nearest lake and recommend self-immersion; that is if you have no horse marine handy.[27]

Our models are very elementary mapping models. Depending on how your brain works, you may or may not find that they help you to visualise the situation. If it does not, then persevere a little, but not for too long. You may well have your own and better way of analysing the problem. And thank you very much, I already know the way to the lake.

ACTIVITY RISK

Now that we have had a first look at the basic terminology and the concept of risk, we can move on from asset risk to the more complicated and interesting problems of activity risk. A central theme of this book is that any activity should be considered in terms of system oper-

27 For the benefit of the under-fifties, 'tell that to the horse marines' is an expression of total disbelief.

ation, whether the activity is boiling an egg, operating a 747, or managing the risks to an asset.

Just as asset risk coexists with an asset, so activity risk coexists with activity. For as long as an activity continues, the risk of an unwanted outcome will exist. For as long as the egg takes to boil, the risk of scalding persists. For as long as an aircraft is flying, it is prone to airborne misadventure. For as long as a ship is at sea, it is at risk to the perils of the seas. The aircraft can avoid airborne risk only by remaining on the ground. The ship can avoid the perils of the seas only by remaining in harbour. The cook can avoid cooking risks by not cooking. However, aircraft on the ground and ships in harbour are not fulfilling their *raison d'etre.* They are not justifying their existence. They must reward their inventors by achieving their design purpose. They must repay their investors by earning their keep. Aircraft must get airborne, ships must put to sea. People must eat. Cooks must cook. Eggs must be boiled. And the fact must be faced that activity risk cannot be eradicated. It is inherent in activity.

The purpose of an operating system is to achieve an intended outcome by controlling the operation. Concomitant with the system activity is the possibility of unwanted outcome. An unwanted outcome translates into the possibility of a loss. The possibility of unwanted outcome may not be high with very simple systems, but it will still exist. It increases with increasing complexity of systems, if only because more components and more interactions will be involved, each carrying some risk of unwanted outcome. Compounding this risk is the increase in opacity which accompanies the increase in complexity. The probability of unwanted outcome can be reduced by skilled risk management, but the risk management measures themselves introduce their own risk factors: their failure, over-reliance on their reliability, and the increase to total system complexity which their presence brings. The purposes of risk management are to reduce risk as far as it is possible, and to minimise the impact of the residual risk.

The possibility of loss accompanies all activities and assets; it can never be eliminated. Risk management measures can reduce risk ever closer to zero, but never to zero. This paradox of risk is a fact of nature, beyond the reach of all management skills. Paradoxically – as you might expect – any deficiency of management skills immediately attracts the waiting legions of risks, full of enthusiasm to serve in their chosen calling.

GENERAL AND SPECIFIC RISK

It is useful to distinguish general risk from specific risk. Polite warning: I have coined these terms to give names to the underlying concepts. They have neither acceptance nor respectability in academic circles, but this does not mean that they are wrong, just that the mills of academe have not ground exceedingly slow upon them. That is exceedingly fine by me and I look forward to the forthcoming flak. Meanwhile we can get on with forthcoming this book.

General risk is a quality inherent in the activity. It is the product of all the risks of the activity, identified and unidentified, internal and external. General risk can be addressed indirectly by applying specific measures to defend against specific risks, and directly by building general safety systems into the operating system, to defend against general and unidentified risks.

A *specific risk* is a quantity, a risk capable of identification. In many areas of activity the specific risks are well understood: the risks of flying by pilots, of the sea by seamen, of patient care by doctors, and of childcare by mothers.

Each specific risk can be addressed and managed. Some can effectively be reduced to zero but, as discussed, the necessary risk reduction measures will add new risks, both specific and general.

NORMAL ACCIDENT THEORIES

One of the great pleasures of studying risk is that there are many excellent, readable, informative books on the subject. Charles Perrow is a sociologist, a profession to the products of which the words readable and informative are seldom attached by non-sociologists. His book

Normal Accidents, first published in 1984, was a major milestone in the development of understanding of risk.

He studied the accident at Three Mile Island, the nuclear power generating plant in Harrisburg, Pennsylvania, USA, which made a serious attempt to go bang in March 1979. This led him to develop a theory which is now known as *Normal Accident Theory*.

In essence, Perrow says that there are two factors which predispose large and complicated systems to accidents.

The first is the degree to which the failure of one component of the system will bring down other components. This knock-on effect Perrow calls *coupling*. He says that the *tightness of the coupling* of components is the first indicator of a system's predisposition to extensive, serious failure. The tighter the coupling, the faster and the further the failure will cascade through the system. If the drain plug on the sump of your car drops out, all your engine oil will be lost. If you do not detect the failure and act smartly, this one failure, the loss of oil, will destroy your engine, cylinder block, pistons, crankshaft, bearings, the lot.

The second predisposing factor Perrow calls the *interactive complexity* of the system. This also is good sense. Each and every component of the system comes with the possibility of its failure. The more components, the more possibilities. And each interaction between each component comes with the possibility of its failure. The more interactions the more possibilities. Add in the fact that the interactions themselves are likely to be highly complicated, often involving computers with their psychotic predilection for bizarre behaviour, and you can see how complexity can quickly head for the infinite. When chemical or nuclear – rather than mechanical – processes are involved, problems are compounded; in such processes it is likely to be both more difficult to see precisely what is going on, and also more difficult to bring them to an immediate halt. As complexity increases, we very rapidly arrive at a point at which we cannot predict all the possible ramifications of a failure. Worse, it is the point at which the poor operator at the sharp end is faced with a situation for which he is untrained and unprepared, and which he cannot comprehend. He will of course have been assured that the occurrence of this state of affairs is absolutely impossible.

Unfortunately the terms 'tightness of coupling' and 'interactive complexity' are terminological flies in the ointment of this otherwise revealing theory. Within each term Perrow places more meaning than the simple words convey.

Tight coupling of components or interactions means that they are bound together; where one goes, the other goes. Or more likely, when one goes, the other goes. There is no buffer – no slack in time or space, nor an isolating switch – between them. The simplest example is of a single component which serves two systems. In some piston engines the oil pump serves as a heat exchanger; the hot oil is cooled by the cold fuel, and the cold fuel is warmed by the hot oil. So if the flow of oil packs up, the fuel will get too cold. Or if the flow of fuel stops, the oil will get too hot. In either case one failure takes out two systems. This is called a common mode failure. It will be followed by another common mode failure; the engine will catch fire and the pilot will get very hot under the collar.

Loose coupling means that there is time or space and the means, perhaps a nice big switch, with which to sever the connection between components or interactions.

If the plug in your motorcar sump is a fraction loose, the oil will leak out slowly. You may see a puddle on the drive, or a warning light on the instrument panel, or see the oil pressure falling; you should have time to switch off and investigate. These interactions are loosely coupled.

If, on the other hand, the plug falls out suddenly, then the oil will be dumped immediately and the engine will self-destruct a couple of seconds later. These interactions are tightly coupled. They will also be tightly coupled to your bank account.

Coupling is, if you like, the insulation between the components and activities of the system. When coupling is loose, insulation is good; shocks are, or can be, stopped. When coupling is tight, insulation is bad and shocks – and thunderbolts – have a free run through the system.

Perrow explains *interactive complexity* by contrasting complex interactions with linear interactions. He says that linear interactions are familiar and expected sequences. He says that complex interactions are unfamiliar or unexpected or unplanned sequences, thereby squeezing three different concepts into the one word 'complex'. One might argue that the concept at the heart of linear is *comprehensible*, and that the concept at the heart of complex is *incomprehensible*. It would be reasonable to hope that the comprehensible would be comprehended. It would be honest to acknowledge that the incomprehensible cannot be comprehended.[28] Unfortunately those who earn their living by designing, selling or directing complex systems are not eager to admit that they cannot comprehend all the possible failures; they prefer to assure us that their systems cannot go wrong. I have heard exactly that assurance given to a room of pilots. It was greeted by a loud cheer and the room promptly emptied. We all went off to the pub for more intelligent conversation. Systems which cannot go wrong have put aeroplanes and their trusting pilots into trees and seas all over the world. It is a law of nature that infallibility is the exclusive preserve of the Gods; humans who presume to enter face not the risk, but the certainty of retribution.

Perrow's Normal Accident Theory says that accidents are normal in tightly coupled (poorly insulated), complex (incomprehensible) systems; the poorer the insulation, the more incomprehensible the system, the more the accidents.

Our *Revisionist Normal Accident Theory* says that accidents are normal in *all* systems. Our theory is based on the risk paradox, the thesis that risk is integral with life, with assets and with activities. Whilst we are agreeing with Perrow that accidents are more likely in poorly insulated, incomprehensible systems, we are going a significant step further. Whether he will agree with us, time will tell. Meanwhile we must remember that in the matter of normal accident theories, our normal is not synonymous with Perrow's normal, and that he got there first.

28 Mr Blair (Perrow's terminology): 'The Prime Minister's role is complex'.
Mr Blair (Malik's terminology): 'I didn't know what I was doing'.

THE QUANTIFICATION OF RISK

We all know that too much salt in the diet is bad for us. In the matter of the quantification of risk, it is too little salt that is bad for us. Claims of precise failure rates, 1 in 10 million, 1 in 100 million, etc., should be taken not with a pinch, but with a bucketful of salt. How do they know the precise figure? Usually they don't. They may be hoping or guessing or misinformed. They may be well-intentioned or they may be engaged in deception. They may have your interests at heart, or they may have their own; the latter is much more likely. The basis of the calculation may be unsound, or the figures themselves may be unsound. If they are estimates of future figures, then they are only estimates. If they are based on historic figures, then they may be accurate, but their applicability to a future period may be optimistic or doubtful or plain wrong.

Unfortunately when we wish to measure or compare things, we find it much easier to do so by using numbers than by other means. In other words, numbers are the mind's default mode for measurement. Fair enough if we are dealing with something which is essentially quantitative or easily quantified. My son runs twice as fast as me; my wife washes up three times faster than me; my wife talks five times as much … . Let's leave it there. Not so good if we are dealing with non-quantities. Rugby gives him twice as much pleasure as chess. Pleasure in what units? We must have units if we are to measure.

And we must measure the right thing. The quickest way of cooking the books is to measure only the bits that give you the result that you want.[29] The Treasury wishes to show an improvement in unemployment figures. Easy. Create a new category: disabled. Move four million unemployed into this new category. Keep quiet about it and, look! We are pleased to announce that we have halved unemployment (sub-text: haven't we done well and aren't we clever?) Answer: no, your figures are crooked and so are you. The Ministry of Defence announces that it has attained this year's target number for

29 As in cooking the verdicts of supposedly objective inquiries. Careful trimming of their terms of reference can exclude all potentially embarrassing areas.

the recruitment of army medical officers. So well done MoD, it has recruited more medics. You must be joking. MoD changed the target to fit the figures. There is more massaging of figures by governments than by whores. Worse, the governments are pleasuring themselves, and not their clients.

In addition to deliberate manipulation of the figures or of their basis, there is the possibility of misinterpretation. In his entertaining and provocative book *Risk*, John Adams of University College, London gives a neat example of confusion. In 1922, 736 children under the age of fifteen were killed in road accidents; in 1986 the figure was 358.

> Allowing for changes in population, the road accident death rate for children is now about half its level of seventy years ago. The child road death rate per motor vehicle has fallen by about 98%.

Objectively, that is statistically, the roads appear to have become much safer. Certainly that is what officials are likely to claim. However surveys suggest a large decrease in children's exposure to road traffic.

> In 1971, for example, 80% of 7- and 8-year-old children in England travelled to school on their own. By 1990 this figure had dropped to 9%.

In other words most parents no longer allow their children to travel to school alone. Their reasons may include crime levels and other factors. However, the parents to some, probably large, degree will be making an intelligent response to the dangers posed by modern high volumes of road traffic: keeping their children away from it. In other words the sands have shifted again; they always do. It is probably because the roads are more dangerous that the statistics make them appear safer.

A final *caveat* relates to the basis of the figures. For example; flying is twice as safe as motoring. Is that per journey or per mile or per hour? Generally or in a specific case? Nationally or internationally, Brighton or Basra? British Airways or Air BraceBrace?

The quantification of risk originated in engineering. A manufacturer of locomotive wheels might find that one in every ten thousand failed in the first year of service, two in every ten thousand in the second year, and so on. In such cases the risk is directly related to wear. The likelihood of failure is a simple function of time, the longer the more frequent. The probability of failure in any given year of an engineering component could be expressed with some accuracy as *n* in ten thousand.

The same reasoning is often applied to modern complex and computerised systems. This is a nonsense. Software tends to fail suddenly and without warning. It does not wear steadily and progressively like a metal wheel. In fact it tends to do the opposite. It starts like a bag of nails and then improves in jumps as the geeks sort out the snags. In 1990 when the Boeing 747–400 was new in service, the software regularly went sick and prevented us, the pilots, from starting the engines. Solution: turn off all electrics, unplug ground power, remembering to forewarn the 400 passengers who will notice the darkness and silence (the forewarning must take place before you plunge them into darkness and silence, because the silence will include the public address system). Five minutes later turn power back on. Software says thank you very much, I'm feeling better now, you may start your engines.

Similarly, complex interactions do not steadily wear out. The unexpected can occur at any time. Perhaps early in the system life because operators have not had time to learn its foibles. Perhaps later in the system life because operators have become less watchful and less wary. So why apply old reasoning to new systems? Because no one can think of a better method of calculating the risk, and no one can think of a better way of expressing risk than in numbers. The method is not totally invalid. It can draw on world-wide statistics for accidents, nuclear, aircraft, or other, so does it offer an indication of probability and risk. Unfortunately precise figures imply precise information, even when the figures themselves are little more than a guess. A physicist and Nobel Prize winner, the late, admirable, Richard Feynman said, 'If a guy tells me that the probability of failure is 1 in 100,000, I know that he's full of crap.' Feynman had investigated the *Challenger* disaster, a lethal failure of risk management, the cause of which had been the use of a material outside its temperature limits. Hence his sharp words. Obviously you and I would not express ourselves in such regrettable terms, but there is nothing to stop us quoting the professional view of a Nobel Prize winner.

UNCERTAINTY AND RISK

The splendidly named Gerd Gigerenzer has written an equally splendid book, *Reckoning With Risk.* Gigerenzer is an eminent academic and research director. The blurb says that the book can help us to overcome our difficulty in thinking about numbers. This it does primarily by dissecting numerical information emanating from supposed experts. In a nutshell, Gerd's message is: don't believe a figure of it. His book is particularly disturbing on the validity of some important medical advice and on the validity of some expert legal evidence.

Gigerenzer makes a distinction between risk and uncertainty. He uses the term *uncertainty* when a lack of empirical evidence makes it impossible or undesirable to assign numbers to the possible alternative outcomes. He uses the term *risk* when an outcome can be expressed as a number such as a probability or frequency on the basis of empirical data. We have already discussed the slightly different definition of risk used by risk management professionals (and ourselves): magnitude multiplied by probability. This difference between Gigerenzer and the professionals is less important than the point on which they agree. As we will see in the next paragraph, both reserve the term risk for situations in which it can be quantified.

Risk management professionals base their definitions of risk and uncertainty on the work of Frank Knight. A seriously clever American economist, he propounded the distinction between the two in his doctoral thesis in 1921. In simple terms:

Risk refers to situations in which the probabilities of the possible outcomes are known, and therefore quantifiable.[30]

Uncertainty refers to situations in which the probabilities of the various outcomes are not known. Most complex situations, and that

30 Be sure to distinguish between probabilities which are known and those which someone claims to know.

must include most situations involving human behaviour, are uncertain. Their outcome is a matter of judgement. Their quantification is therefore not a calculation but a speculation. It may also be a deception. Whatever, don't bet on it.

We may relish the thought of the merry battles which must take place over which of the terms, risk or uncertainty, is the more appropriate in any given case. What is important to us is that both Gigerenzer and the professionals reserve the word 'risk' for cases in which the probabilities are truly quantifiable. We may strongly suspect that Gigerenzer would need much more convincing of the case for quantification than would others. We would most certainly side with him. Let us resolve to greet all claims of precise probabilities with scepticism and a bucket of salt. If pressed we can cite the professional opinion of a Nobel Prize winner.

PERCEPTIONS OF RISK

When people get their knickers in a twist, it is not their ambulation which is impaired, but their cogitation. It is a situation which most of us take pains to avoid. However, if ever you find this state of affairs creeping up on you, you are in good company. In 1992 the Royal Society invited a group of eminents to form a Working Group to conduct the Royal Society's second study of risk. The group comprised engineers and social scientists. The engineers espoused the quantification of risk; the social scientists took the view that risk was not a matter of quantification but of perception. A merry and doubtless highbrow ding-dong took place. There was no winner and the two parties were unable to find common ground. The Royal Society, presumably chagrined at its inability to produce an *ex cathedra* viewpoint, published their findings, not as a Report of the Royal Society, but as a Report of the Working Group. The two differing viewpoints were presented separately in the report. Boxers versus Y-fronts. We are in good company.[31]

31 When I reread this first paragraph, it came to my notice that the 'i' and the 'o' of my qwerty keyboard were adjacent. In typing the fifth word, I had hit the wrong one, thereby

The social scientists' view, that risk is a matter of perception, means that risk is what a person thinks that it is. Remember that risk is a man-made concept; you cannot touch it or handle it; you calculate it or feel it. Thus far we can agree with the social scientists. The final measurement of – and decision on – risk is in the mind; it is a mental process.

On the other hand, one important determinant of the quality of the decision is the input into the mind, the information on risk which informs the consequent decision. The mind is seeking to measure the risk, perhaps to compare two risks and judge between them. The mind finds measuring and comparing different qualities difficult. It finds measuring and comparing numbers easy; it likes to quantify. It will probably start by applying a common scale of units to each quality. The scale most readily available is units of money. If money is clearly an inappropriate or unacceptable measure, then a different unit must be devised: units of happiness, of satisfaction, of thrill, etc.

We thus find ourselves in the unusual position of agreeing with both parties. It is indisputable that the risk decision is a mental process, so hurrah for the social scientists. It is indisputable that some risks can be quantified, albeit only to a certain degree of accuracy, so hurrah for the engineers. Simple quantification may indeed give the answer to a question of pure engineering risk (cantilever or suspension bridge, welds or rivets). However once beyond pure engineering the quantified risk is likely to be only one of several risk factors: cost, time, reliability, logistics, policy, etc. At this point the quantified desirability of rivets must be assessed against all the other factors, and we are back to the problem of comparing unlikes. Without a common scale we cannot decide risk by means of quantification.

Our conclusions on quantification are:

- Risk may be quantitative or qualitative
- Quantification can be an objective process

coining a phrase unlikely to appeal to feminists. I mention this only to save you from the same mistake, and to show that Murphy rules everywhere, including in books on risk.

- Quantitative risk can be measured by quantification, if and when that quantification is valid
- Quantitative risk measurement is valid only in strictly defined situations. In all other situations, risk is qualitative
- Qualitative risk is subjective; it is a matter of perception
- If others seek to use quantification to influence your qualitative risk judgements, keep salt handy
- If the government seeks to use quantification to influence your qualitative risk judgements, keep two fingers handy

CONCLUSIONS

1. A loss is the removal of the benefits of the asset
2. Risk, by widely accepted definition, is the: *Magnitude of possible loss multiplied by the probability of occurrence of loss*
3. A simple model of asset risk management comprises asset, loss, hazard, stakeholder and risk
4. Separate models are needed for each stakeholder
5. *General risk* is inherent in any activity
6. *Specific risk* is a risk capable of identification
7. Our *Revisionist Normal Accident Theory* states that accidents are normal in all systems
8. The quantification of risk is a mathematical process
9. The perception of risk is a mental process
10. The quantification of future risk involves assumptions. It may incorporate error or deception or both
11. The results of quantification should not be taken without salt
12. The results of most government quantification should not be taken at all

CHAPTER 4

OF CAUSES AND CANARDS

WHERE HAVE WE BEEN?

This is the fourth and last chapter of Part I. Our first three chapters explored some important concepts. If you skipped some of the heavier paragraphs, then do the same to this; speed read and get the gist of the essentials.

In Part II we move on to easier stuff. We examine real-life accidents and discuss how they happened and how they could have been prevented.

By the end of the book I very much hope that your interest in accident avoidance will have been aroused. If so, have another go at the concepts in Part I. With heightened interest you will easily absorb them, heavy bits and all.

WHERE ARE WE GOING?

To prevent accidents we must first understand them. In particular, the true natures of their causes are very important to us. We know from experience that, given half a chance, just about anything in life can go wrong. We also know why. If we offer the Gods an opportunity to zap us, they may well yield to the temptation. Unfortunately the only lesson in prevention which we can draw from this is that we should deny zapping opportunities to the Gods. This is an undoubtedly sound as a fundamental policy, but it is somewhat weak on the finer detail of accident causation. We need a comprehensive understanding both of its generalities and of its specifics. We can then set about reducing, and when possible eliminating, known causes. This will oblige the Gods to think up some new ones.

In this chapter we will seek to identify the concept encapsulated in the term *accident cause.* First we discuss the need to prioritise accidents, to sort the real nasties from the trivia. We look at hindsight, a concept which sometimes illuminates and sometimes confuses. We review two different theories of cause. The first is the *Evolutionary Theory* – the chain of events view that the system defences tear down a dotted line of chronological events to culminate in an accident. The second is the *Big Bang Theory* – the reductionist view of root cause analysis that an accident can be traced to a major underlying cause or causes. These two views are fundamentally incompatible. Happily we do not have to engage in a protracted analysis to determine which is right. The evidence is that both are wrong, albeit to different degrees.

Finally we get off the fence to put our money where our word processor mouth is, and spell out a different view of accident causation.[32]

NASTIES AND TRIVIA

Careful definitions may be an essential preliminary to a careful discussion. However, the passion aroused by disputes on definitions very often thickens into an acrimonious pea-souper, from which emerges neither definition nor careful discussion. We do not wish to be so befogged; therefore let us call an accident an unwanted loss; I know what I mean by unwanted loss: you know what you mean by unwanted loss. It does not matter that we may not agree on the specifics of the loss. What does matter is our broad agreement that an unwanted loss is something which we try to avoid. There is our definition: *an accident is an unwanted loss.* Full stop. Pause while you think about it; please try to avoid mother-in-law jokes. And if an academic challenges this definition and asks you how you justify it, tell him/her that you argue it on the grounds of common-sense. As we know, academics do not accept the concept of common-sense, so they debar

32 Sitting on the fence is said to lead to a posterior resembling a hot cross bun. The transverse groove being the impression left by the fence. No one would wish to be called a hot cross bum.

themselves from our battleground. Sad – boo-hoo. But we can proceed with our discussion – hurrah.

Both the number and the range of potential losses with which we are faced are vast. Anyone we know can suffer harm. Anything we do, anything we own, can suffer mishap or loss. Our resources are finite; a resource devoted to defending against one loss is not available to defend against another. For that reason it is vital to prioritise our potential losses, to pick out the real nasties from the vast clutter. We can then concentrate our resources where they are needed. Equally important, we can avoid frittering them away on trivia.

We need a basis on which we can determine the significance of accidents or adverse events. Then we can rank them in order of significance, and allocate our resources appropriately. Let us look at some different activities. We will use the term *extreme adverse event* to describe the most serious outcomes. If we were a mother in a kitchen, an instance of an extreme adverse event would be the burning of a toddler by a hot liquid. If we were a simple non-hazardous business, an extreme adverse event would be bankruptcy. If we were an airline, an extreme adverse event would be an air accident. If we were a nation, an extreme adverse event would be engaging in an unnecessary and unjustified war and an unwinnable occupation, causing hundreds of thousands of deaths, including those of many of our servicemen. (You would think. But then you were not Prime Minister.) From these examples we can see that there is no single unit common to all activities. There is therefore no single scale along which we can rank adverse events in order of importance. A unit and scale appropriate to the context must be determined for each situation.

THE IMPORTANT DIFFERENCE BETWEEN HINDSIGHT AND FORESIGHT

Hindsight is an unfortunate word; it appears to mean foresight. 'With hindsight, I would not have done that' equates to, 'with foresight, I would not have done that'. Confused? Join the club.

To leave the club all you have to do is to remember that hindsight is wisdom after the event. If you look back from an accident over the period of time which preceded it, the line connecting the accident to some past event X is often clear. Tracing this line back is finding out what happened; it is the essential first step in investigation. It allows the past event X to be identified as a possible accident cause. We may hope that this, in turn, will enable us to prevent future accidents from the same cause.

Take a motor accident. If we find that the driver lost control, that this was caused by a tyre burst, and that the tyre was defective, then the fitting of this tyre is event X. We can take steps to prevent the fitting of defective tyres.

If, however, our task goes beyond investigation and we are trying to establish responsibility and possibly blame, wisdom after the accident goes out of the equation. We must put ourselves back to the time of event X. The question is then of foresight, whether with the knowledge available at this time, the line connecting event X with the accident was reasonably foreseeable. Often the possibility of an accident will indeed have been clearly foreseeable by someone. In most cases foreseeability will be greatly disputed. Was the tyre fitter aware that the tyre was or might have been defective, that its fitting might lead to an accident? Should he have known?

The past is known. It is a single chain of identifiable events. The future is unknown. It holds a multitude of possible events and an even greater multitude of possible chains of events.

Consider a large tree. How did it grow to that exact shape? Easy. Here are snapshots of it going back over the years and here is an example of the seed from which it sprang. Wisdom after the event. Fine.

Now try this. Here is a seed. Its variety is unknown. Decide whether it is poppy or Californian redwood or one other of a million possibilities. Now please describe its future life cycle, all its possible experiences, precisely how it will grow, which branches will fall off, and its exact final shape. No pressure. But bear in mind that your any failure of prediction will, if at all possible, be found to be your fault. And you will be fired. Or locked up. Or dead.

SEED

Actually the seed might go into your breakfast muesli and not grow at all. Equally event X could have gone nowhere; it could have been cancelled out seconds later by event Y.

In accident prevention we must use both hindsight and foresight. Hindsight to incorporate learning from past events into new and safer practices. Foresight to foresee which of today's seeds will, if unchecked, grow into tomorrow's accidents. You would expect foresight to present difficulties, and you would be right. You would expect learning from past events to present few difficulties; many excellent books have been written to show that you would be horribly wrong (see endnotes for list). In fact a powerful case can be made that every single accident which occurs will have had a forerunner; it will have been presaged by another event, the lessons of which had not been learned. From fatal aircraft accidents and from fatal medical accidents and despite unprecedented litigation awards, bad practice emerges and survives, free to cause further accidents.

CHAIN OF EVENTS: EVOLUTIONARY THEORY OF CAUSE

The 'chain of events' model is the *Evolutionary Theory* of accident causation. It is helpful in some ways but misleading in others. It is helpful in that it is conceptually similar to a crack which spreads event by event through a structure until some final event completes the fracture and the whole structure fails. This encapsulates the concepts of a structure, of the connection of a number of entirely separate events to form a chain, and finally of an accident as the product of that chain and not as the product of a single flawed action by the final operator.

This last concept, that there is always more to accident causation than operator error (yes, always, as you will see) is accepted by almost all academics and professionals engaged in risk management. Would that it were equally accepted by Air Chief Marshals, chief executives and other hierarchies whose performance comes into the firing-line if the investigation is not conveniently stopped by a verdict of operator error.

The late Barry Turner pioneered the modern approach to accident analysis in his book *Man-Made Disasters*. When it was first published in 1978, it fell fairly flat. Interestingly his thesis was established in mainstream thinking not by peer group appreciation but by a nudge from the Gods. It was a series of disasters, Three Mile Island (1979), Bhopal (1984), Chernobyl (1986), *Challenger* (1986), *Herald of Free Enterprise* (1987) which graphically demonstrated Turner's insights. His thesis, in a nutshell, was that a disaster was not an Act of God but the culmination of a process. Turner chose to analyse this process on a time base, dividing it into stages: starting points, incubation period, precipitating event, and so on. This provided a foundation for the concept of disaster as the product of an extended process, and for differentiating that process into its constituent events. It put the boot firmly into two earlier concepts, that of disaster as simple mischance or Act of God, and that of disaster as the result of nothing but operator error. For this latter contribution alone, pilots and other final operators the world over should raise a glass to Barry Turner. Organisational hierarchies can no longer close down all inquiry by finding that their own management was faultless, that operator error was the sole cause, and by firing the said operator. They still try, but now they run into years of flak which knocks chunks off their reputations. And serve them right.

The links in the chain of 'events' may be of many different forms. They may be human: beliefs, advice, customs, actions, etc. They may be technical: components, processes, materials, etc. They may be bureaucratic: directions, orders, procedures. They may be occurrences, internal or external.

In selecting a time base for his analysis, Turner highlighted the fact that many of the 'events' which lead to the accident, precede its occurrence. They lie in wait, links in the chain, lengths in the crack, until some final event completes the chain, and the whole structure is fractured. They are commonly referred to as latent errors. Errors may be something of a misnomer. Some of the links may be perfectly proper actions in a normal course of events. However, if the structure fails, if they form links in the chain, if the crack passes through them, then they become errors in the context of that failure. They may not be errors in any other context. Is it fair to call them errors? Yes and no. But actually yes. They may have been perfectly proper in most contexts but they failed to act as crackstoppers in the context of the accident. Therefore yes, they were complicit in the failure of the structure.

THE FATAL ROW

The chain of events model is to some degree misleading because it implies a chronology of events which follow one after the other, a steadily lengthening chain which culminates in an accident. A better picture is of a system in which small cracks are appearing. They do not occur in one place, each successively lengthening the original crack until it is complete. They appear in different places. They may be vertical, horizontal or diagonal. They occur in random order. These small cracks are a combination of events, conditions, deficiencies and errors. They gradually accumulate and spread across the system, deep-rooted crazing. There they remain, latent, awaiting only one final event, one action or mischance to connect all the small cracks together into a single complete fracture. If at this point such a final event occurs, lightning strikes, latent errors become active errors, all system defences are breached, and an accident occurs.

Think of a fruit machine. Only a complete row of a particular fruit will give you a jackpot. Each winning piece of fruit pops up in no particular order and remains in place in the row. Finally only one more piece is needed to complete your jackpot row. It comes up. You win. Yippee. You note that each and every piece of fruit is essential to the desired outcome.

Accident aetiology (causation) is the same, except for two things. First, it is latent errors which pop up and remain in place. Second, we desperately do not want to complete the row because, if we do, the outcome is not a jackpot row. It is a *fatal row*. An accident.

We note again that, without each and every item, the fatal row would not be complete; the accident would not occur. For that reason, each and every item in a fatal row is an accident cause. The plethora of operator error verdicts is the result of attributing accidents only to the final event, the one that completes the fatal row. Any such verdict ignores the other items in the row. It is simply wrong. It is not a verdict which could be pronounced by anyone with the most elementary understanding of accident aetiology.

HUMAN ERROR (PLEASE DON'T GO AWAY)

On the one hand, human error is a vast subject, as difficult to pin down as human behaviour itself. Sorting the wheat from the chaff in the large libraries and interminable conferences devoted to the topic is singularly time-consuming and unrewarding. We are not going to attempt it. Therefore please don't go away.

On the other hand, as an accident verdict, human error is a soundbite which is easily understood. It allocates responsibility and focuses blame. It fulfils the need to remove uncertainty because it provides an answer. It may often be a half-truth, thus it may carry more conviction than it merits.

There is a strong argument that human error is the fundamental cause of every accident. If a structure fails: aeroplane, bridge, whatever, someone somewhere will have chosen its material, specified its strength, designed its form, overseen its construction, maintained its condition, directed its use, applied or allowed the application of its load, monitored its performance, controlled its operation, and failed to prevent its failure. Human factors and potential human errors are at every juncture. Even in the case of new materials or new technology where the failure could not have been predicted, someone will have made the decision to utilise that new and relatively untested material or technology. So why, one wonders, does the term human error appear to be reserved for final operator error? Official inquiries often find operator error. But they do not find prime ministerial error, ministerial error, chief executive error, Air Chief Marshal error. Infallibility must be one of the privileges of rank, and human error a failing of the lower orders, confined to those below the salt.[33]

33 Below the salt: a bad seat at the dinner table of life.

As we will see, investigation of accident cause is the question 'why?' repeated until a satisfactory answer is achieved. What constitutes satisfactory? If the overriding purpose of the investigation is to prevent repetition, the only satisfactory answer is the identification of a cause that we can eliminate or correct. Human error most certainly cannot be eliminated, and nor is it a failing sufficiently specific to enable its correction. For that reason (let alone all the others), human error can never be an acceptable, satisfactory finding of an accident investigation.

In fact, human error is not a cause. It is a consequence of a condition, whether of the situation, or of the operator's selection or training or mental state, or other.

There are various ways in which the generality 'human error' can be differentiated until specifics are identified. Professor James Reason is one of the best-known proponents of this approach. More important, he has a record of practical achievement with his methods. He first differentiates error into:

- Slips (attentional or perceptual failures)
- Lapses (failures of memory)
- Rule-based mistakes (rules misapplied or bad rules or good rules not applied)
- Knowledge-based mistakes (no standard operating procedure for the situation), and
- Violations (deliberate or inadvertent deviation from rules)

We can see that with this first differential the good Professor has already identified one cause, the fourth, no standard operating procedure, and has made good progress with the others. That makes our first point, which is that in investigation we must continue to ask why until we identify a cause which we can correct. So we have no need to dive deeper into differentials (phew).

However! To make our second point, we must examine two more sets of Professor Reason's differentials (the last, I promise). In the early 1990s he was invited by British Airways to assess the organisational and local factors relating to error management in aircraft maintenance at Heathrow. He listed twelve local factors. Below, to give you the flavour, are six of them:

- Knowledge, skills and experience
- Morale
- Tools, equipment and parts
- Paperwork, manuals and procedures
- Fatigue
- Pressure

And here are four of his eight organisational factors:

- People management
- Provision and quality of tools and equipment
- Training and selection
- Commercial and operational pressures

Have a look at these lists. They describe the situation in which the maintenance workers find themselves. By and large this situation is imposed by the management, with little input from the workers. The organisational factors are the background from which the local factors arise. The local factors are the background from which human error arises. The organisational and local factors are therefore a possible, even probable, cause of any human error.

Contrast Reason's analytical approach with facile verdicts of operator error. His investigation does not stop with human error; it starts from it. He burrows down for its detail. He asks 'why?' He looks back into its situational background for its underlying causes.

If we were to criticise Reason, it would be for his easy application of the term human error to operator error, and his failure to apply the same term to the management error which is manifest in the failings of the organisational and local factors.

SITUATIONAL BACKGROUND TO ERROR

The situational background to error is a big subject about which many weighty tomes have been written. We are not going to add to their number. What we will do is to give life to a rather ponderous phrase by looking at some examples.

Consider stress in general. A little stress is good; it keeps you on your toes. Too much stress is bad, it scrambles the brain; thinking calmly and acting deliberately become difficult. Take three situational backgrounds. Carry that empty cup to the kitchen. Easy. Carry that very full mug of hot coffee across your wife's new carpet. Hmm. Carry that very full glass of liquid high explosive across that icy motorway. No, thank you.

Consider time pressure, perhaps the simplest form of stress. The faster that you have to work, the less time that you have to give careful consideration to each action, and the greater the probability of error or omission. In simple tasks, you can commit the required sequence to memory and then carry it out automatically. In an unfamiliar task this is obviously not possible. In complex or potentially hazardous tasks it is highly undesirable; automatic actions do not provide the care and deliberation which is vital to the avoidance of error. For such care and deliberation, time is needed. If it is not allowed, an operator is under time pressure. He is likely to make a slip, a lapse, or a rule-based mistake. Consider your own experience. How many times have you said, 'you wouldn't have done that stupid (whatever) if you hadn't been in such a b—— rush'? Road accident – how often was someone in a hurry? Reports of major accidents give many examples of operations under time pressure: *Titanic*, R101, *Challenger*, Tenerife, and lots more. These accidents are discussed in the next two chapters.

One of the arts of operating a complex aircraft is to manage time pressure out of the equation as far as is possible. Give yourself time to get everything right, to complete your checks, to think. Meticulous adherence to checklists is one of the bedrocks of flight safety. As a young flying instructor in the RAF I used to say to students, 'Right, now for some aerobatics. Don't bother with the pre-aerobatic checks, we are tight for time'. 'Yes, sir' is the only wrong answer to this improper instruction. There are as many right answers as there are ways of saying 'no, we will do our pre-aerobatic checks and then we will do our aerobatics'.[34] Unfortunately in unprofessional aviation circles, that is outside decent airlines and well-run units in the forces, there are the brainless who consider it macho to blast off, never mind the checks, we are so good that we don't need a checklist.[35] Read about them in accident reports and on tombstones.

Workplace design is a part of the situational background. An aircraft has levers for the undercarriage, the flaps and the airbrakes. It is good to select the right one. It is therefore good to be able to tell them apart. It has instruments to convey important information; it is good to be able to read them easily and surely. The controls are better now, but for decades aviation manufacturers in Britain and the USA designed cockpits, if not for prettiness, then for designers and not for pilots. They grouped undercarriage, flap and airbrakes levers neatly together. Guess what? Errors were designed into their situation, accidents were waiting to happen. And they did. Some of them are described in the next two chapters.

Instruments were as bad. Pilots trying to decipher altimeters with three needles, one each for 10,000s, 1000s and 100s of feet, regularly misread them and walloped into the ground which they thought was 10,000 feet below. The crew of the British Midland Boeing 737 which crashed at Kegworth in 1989 had their task of dealing with an engine failure compounded by both the infelicitous design and the known faults of some of their engine instruments. 47 died. Airbuses too numerous to list have crashed because of situational factors in their clever new cockpits.

Andrew Brookes, whose books on air accidents are always worth reading, quotes my friend Roger Green, who was Head of Psychology at the RAF's Institute of Aviation Medicine.

> If you give [a pilot] training that doesn't equip him to deal with a situation, are you going to call it pilot error if he gets it wrong? And

34 The answer which I used to recommend over the more pungent responses was, 'no, sir, I am finishing my Murraymint, my too-good-to-hurry-mint', a contemporary witticism which will mean nothing to the under-forties. Young upstarts.

35 The Germans have a nice word for the brainless: 'halbstark', halfstrong. Used for those whose body has grown, but whose brain has not.

if you give him equipment which isn't as easy to use as it could have been, are you going to blame the pilot who maybe gets it wrong when he doesn't get time to sort it out?

Sadly, Roger Green is no longer with us. Amongst his many admirable contributions to good sense and to aviation safety is the anonymous incident reporting system, CHIRP. [36]

The objective of accident investigation is accident prevention. No investigation is complete until it has found a cause which can be corrected. Years of findings of human error have been followed by yet more years of findings of human error. Prevention has not been achieved. Inquiries and investigations which end with the finding of human error are not fulfilling their responsibility.

THE FINAL OPERATOR

Pilots, it is said, are the first on the scene of the accident.

Contrast their presence with the likely absence of those responsible for aircraft design and manufacture, with the likely absence of the directors or Air Chief Marshals responsible for choosing the aircraft, with the likely absence of the flight management or Air Chief Marshals responsible for ordering the operation, with the likely absence of those responsible for writing or providing, or not providing, their operating limits, flying manuals, handling notes and checklists, with the likely absence of the supervisors and technicians responsible for the maintenance of the aircraft. Be that as it may, there are three good reasons for investigating the final operator's role.

First, by the nature of his role, an action of the final operator will often be one of the items in the fatal row, perhaps the final event which completed the row. It was the pilot at Kegworth who misunderstood his infelicitous situation. It was the pilot of the Trident who

36 There are two Roger Greens who have contributed mightily to aviation. The second is Dr Roger Green, formerly a senior aviation doctor at British Airways. With others they co-authored a book, *Human Factors for Pilots.* Dr Roger told me that a friend, seeing the two Rogers together, each of distinctive appearance, remarked 'If I were God and making a Roger Green, I would not have made him look like either of you.'

raised the leading-edge flaps and caused the aircraft to fall out of the sky (1972).

Second, again by the nature of his presence, the final operator may be the only person in a position to have averted the accident. He is the goalkeeper. When all other defences have cracked, only he remains; he is the crackstopper of last resort.

Third, it may be that the primary task of the final operator, his highest priority, is to avoid or avert an accident. This is certainly the case in civil aviation. Passengers wish to get down safely; where or when is of far less importance; bonus points to the pilot for right place and right time.

There are thus three good reasons for scrutinising the role of the final operator in any accident. There are a lot of very bad reasons for placing all responsibility and all blame upon him. The two worst of these very bad reasons are the dreadful injustice of placing blame where it does not belong, probably ruining a reputation and a life, and the concomitant failure to address and correct the true accident causes.

The answer to the conflict between the good reasons and the bad reasons is competence and integrity in investigation. That means the compilation of adequate *information*, the *intelligence* to understand it, and *integrity* in its analysis. Many self-serving hierarchies find this answer unpalatable.[37]

You may be beginning to feel that this book indulges in special pleading for pilots and other final operators. Yes, and then again, no. Yes, I most certainly do plead for information, intelligence and integrity in investigation. No, I most certainly do not plead for the exoneration of the incompetent or professionally inadequate. Get them up to standard or get them out. No one, and certainly no pilot, would deny that a pilot can himself be the cause of an accident. The difference to be explored is that between unforced error and forced error, that is to say error forced upon him by his situation. The important question is

37 Note these three i's, information, intelligence and integrity. We will meet them again whenever we discuss decision-making.

how other pilots would have acted in the same situation; the more probable it is that they would have acted similarly, the more the probable it is that the error was forced. Exactly the same reasoning applies to other professions: soldiers, sailors, nuclear power plant and chemical process operators, *et al.* All investigations of final operator performance must start with the situation in which he found himself. And what and who put him there. Safe operation is the product of a good system and good management. The final operator is there to prevent the accident planned by the management.

ROOT CAUSE ANALYSIS: BIG BANG THEORY OF CAUSE

Root cause analysis is the *Big Bang Theory* of causation. At its simplest it postulates a single underlying cause. Root cause analysis is not a helpful term because it is applied to widely differing methods of analysis. Most are proprietary and directed at the analysis of business process failure, not at the analysis of accident causation. They are all structured to a greater or lesser degree. They all seek to identify a root cause or causes, but they do not share a clear common definition of root cause. Some define the root cause as the first cause which can be both identified and addressed. Others do not offer a specific definition, but look beyond a single cause. They trawl through the system noting those causes which they deem important. Some methods of root cause analysis may be useful. Their value depends upon their definition of root cause and their methodology.

Most methodologies make repeated use of the question Why? They emphasise the need to distinguish cause from effect, making the important point that it is determined by viewpoint; from one direction you see cause, from the other you see effect. Looking forward from the patch of oil on the floor, it was the *cause* of my aerobatics in the garage. Looking backward from the patch of oil, it was the *effect* of the leaking sump of my elderly MG. Looking backward from the leaking sump, it was the *effect* of the quality of workmanship in the British Motor Corporation. Looking forward from my aerobatics, they were the *cause* of the mirth of my sons. The mirth of my sons should have been the *cause* of a good kicking; sadly it was not; they are both bigger than me.

You can see how this process quickly builds an event chain. It can continue, if not *ad infinitum*, then certainly *ad nauseam*. The problem is to know when to stop. One eminent writer on accidents, Rasmussen, says that there is no formal stop-rule. Instead he suggests three pragmatic solutions:

- When the causal chain comes to a dead-end for lack of information
- When a familiar abnormal event is found to be a reasonable explanation
- When a cure is available

Rasmussen is right to say that there is no fundamental principle to tell us when to stop (or at least, none has yet been found). None of his three solutions is entirely satisfactory. In the context of accident prevention, the last is the best. However, if we stop at the first cause which can be both identified and addressed, this may well leave other latent errors in place, and some other accident waiting to happen. We must do better than that, and we will in Chapter 8.

Root cause analysis is a reductionist methodology, one which seeks to boil down or reduce something to its simplest constituent parts. In this case the analysis goes on boiling down until it finds what it calls a root cause. But who says what constitutes a root cause? Worse, the concept of a root cause is itself dubious. The eminent geneticist, Professor Lewontin of Harvard University, calls the concept of simple unitary causes an ideology; this means an unscientific belief. He emphasises the need to distinguish agents from causes. Tuberculosis comes from the tubercle bacillus which in turn thrives on the undernourished and which in turn spreads in sweatshops which in turn are the product of dynamic capitalism. Where does the music stop? Which is the root cause? Lewontin argues that asbestos and fibres are agents, not causes, of cancer. Eliminate them and you do not eliminate cancer, other irritants will take their place. Asbestos and fibres are agents of the social environment and it is here that some deeper causes may lie.

Confusion arising from the somewhat complicated concepts of cause and agent is understandable. Less forgivable are the deliberate misidentifications of root cause. The former Chancellor of the Exchequer told us that he was the root cause of our prosperity because he controlled the economy. Whereas, he now tells us, our present straightened circumstances are not caused by him, but by matters outside his control. I must call the station-master at Milan.[38]

Ignorant misinterpretation of root cause often suits the misinterpreters' convenience. For example, those responsible for museums have been known to claim every museum visitor as a cultural tourist. Each tourist spends £large on his holiday; therefore the museum is the cause of a contribution to the national economy amounting to £large multiplied by number of museum visitors; therefore please increase the museum's budget. Oh, yes? It started to rain, museum entrance was free, the visitor popped in to keep dry. Cultural tourist? Rubbish.

View all claims of the identification of a root cause with a degree of scepticism. There be ignorance and vested interests out there.

There is no scientifically-based definition of root cause. Root cause analysis is a term which is regularly applied to any method claiming to get to the bottom of things. It means different things to different people, if indeed it means anything at all.

CAUSATION

Our first model of causation is the crack in the operating system. Our second model is of this crack as a path through system defences; we call it a fatal row. Our models are not dissimilar to the chain of events

38 Ian Ponsford, a great man and a great friend, was a wartime Spitfire pilot and eminent lawyer. When confronted with an obtuse or obstructive person, he would remark, 'I must call the station-master at Milan'. If that person unwisely asked for an explanation, he got it. 'I told the station-master at Milan that I would call him immediately, if I ever met a bigger bloody fool than him'.

model, but with differences discussed above. Let us look only at what we mean by cause. Pause here and, if you wish, try to define it.

Alexander McCall Smith touches on the subject in his splendidly readable book *Espresso Tales.*[39] Paraphrased, here are his two examples. Small boy, aged 5, stands to inherit a vast fortune. In the event of his death the fortune goes to his wicked uncle. Wicked uncle arrives at house, finds it empty except for small boy. Small boy is in bath.

Case A. Wicked uncle drowns small boy.

Case B. Wicked uncle sees small boy in bath, drowning. Leaves him there.

Case A is straightforward. Is uncle the cause of death in Case B?[40]

Some bright academics have suggested that there is a difference between the scientific and legal concepts of cause. The scientific concept, which they call causality, is that when A follows B *as a direct result* of B, then B is the cause of A. The legal concept, which they call causation, is that when A would not have occurred *but for* B, then B is a cause of A. One problem with the scientific concept is the definition of the word 'direct', and consequently of 'indirect'.

The concept of causation by no means solves the law's problems with cause. The basic test of negligence is whether, on the balance of probabilities, the loss would not have occurred but for the actions of the defendant. This can pose difficulties. Where the loss is a loss of health, there may be many possible causes. Proving the case against any one particular defendant is therefore at best tricky, at worst impossible. The House of Lords' solution was to move the goalposts. Instead of the established test (but for the actions etc., above), the Lords set a different test, whether the actions of the defendant increased the probability of the plaintiff's loss of health.

Where there are several possible sources of the loss, no single one of which can be identified as the cause in any given case, another solu-

39 Mention of which reminds me. Let us raise a glass to Bertie who, aged 6, set fire to his father's *Guardian*. A much greater contribution to society than burning unreadable avant garde novels. We need more Berties.

40 In both cases the uncle discourages small boys from washing, a capital crime and enough for most judges.

tion has been adopted. They are deemed to be jointly and severally liable. Leukaemia, for example, might be the result of pollution from any one of several industrial plants. Joint and several means that any one of them may be sued and held liable for the loss. In each of these examples, the impossibility of pinpointing the cause would have left victims uncompensated, and offenders untouchable. The law did not attempt the thankless task of redefining cause. Instead in one case it moved the burden of proof from causation to complicity, and in the other it spread the focus of liability. The nature of cause is not as simple as it sounds.

Let us return to the concept of a fatal row. Assume for the moment that each item in the row is equally important, and that the removal of any one item would certainly (not probably, possibly or maybe) have prevented the accident. It follows that each item in the row offers an opportunity (by its removal) to have prevented the accident.

Our objective, our perspective, is the prevention of accidents. From this perspective, any failure to prevent the accident is an accident cause. Each item in the fatal row is an accident cause because its removal would have prevented the accident. And there is our definition of cause. *Any failure to take an opportunity to prevent an accident is an accident cause.*

Clearly the question of opportunity is paramount. Further, the issue of duty of care is central to any assessment of responsibility or blame. It would determine the outcome of Regina v. Wicked Uncle, Case B, above.

Our definition satisfies both the scientific and legal concepts of cause. As A accident follows B failure of prevention, then B failure of prevention is the cause of A accident. As A accident would not have occurred but for B failure of prevention, then B failure of prevention is the cause of A accident.

Our definition is also applicable in the cases of multiple possible causes discussed above. For example, it is not possible to place upon an industrial plant the responsibility for preventing leukaemia in local residents. However it is possible to require the plant to take every opportunity to prevent its activities from increasing the probability of leukaemia. If the firm failed to take every opportunity for prevention then, by the standard laid down by the Lords, it would be deemed to have increased the probability of leukaemia, and would be found to be liable. Thus by our definition, the firm is a cause of the leukaemia. Note that we are not saying that it is the only cause.

Our definition leads into the final topic of this chapter.

THE SYSTEM OF OPERATION

We have already suggested that any activity, however simple, can be analysed as a system of operation. From that perspective, all systems of operation must have three primary objectives, Types A, B and C:

- Type A objective – conduct of the operation
- Type B objective – avoidance of extreme adverse events
- Type C objective – conformity with regulations

To meet objective A, mother must boil the egg, the aircraft must fly or the ship must go to sea.

To meet objective B, the toddler in the kitchen, the aircraft or the ship must not suffer harm or loss; we must not have an accident.

To meet objective C, the operation must conform to regulations.

The system should be designed to achieve all three objectives, that is, to manage their associated risks:

- Type A operational risks – the risks to the smooth running of the operation
- Type B extreme risks – the risks of an extreme adverse event
- Type C regulatory risks – the risks which arise from statutory law, official and semi-official regulations, guidance, codes of practice, advice, etc., which may be promulgated by HMG, by the European Commission, or by any of the Mongol hordes of nosy parkers and busybodies who beset modern life. There is no body regulating the activity of boiling an egg. Yet.

Types A and C risks are the basic risks of business. Those who cannot handle them should not be in business, and probably soon will not be. Types A and C risks should be addressed by good business practice, perhaps with a posh name like Total Quality Management or ISO something or other. Type B risks, the risks of extreme adverse events, are the main subject of this book. They should be addressed by a programme of active prevention, specific to their context. They are the risks which matter to us.

Note that in our terminology the risks to financial businesses which manage investments and insurance are Type A risks. They are the normal operational risks of their chosen activity.

The problem is that in many, perhaps most cases, Type B risks will tend to be subordinated to Types A and C. In commerce, failure to meet objective A, the smooth running of the business, will lead to loss of profits and to bankruptcy. Failure to meet objective C, conformity with regulations, is likely to meet with immediate retribution from *les miserables*, the massed ranks of regulators, the sole fulfilment of whose empty lives is the enforcement of conformity. Often Type C risk may be the most immediate risk to organisations and, more important, to their principals. Type A risk is likely to be, if not the most immediate, certainly the biggest risk to the principals. A substantial financial loss will cost the chief executive his job. A major breach of regulation may cost him his job. One accident probably will not. Money matters to him. Regulations matter to him. Let us not say that accidents do not matter to him. It is just that it is in the nature of things that cash-flows and regulation changes will be at the top of his in-tray, and safety and Type B risk below them.

For the moment, that is where we will leave the subject of the operating system. The main proposition is that the avoidance of extreme adverse events is one of the three prime objectives of the system. This proposition reinforces the concept of an accident as a system failure.

It is an unfortunate fact of life that hierarchies are not judged on their organisation's safety record. Unless the record becomes a disgrace; then it is too late.

CAUSES: OUR CONCLUSIONS

1 By our definition, an accident is an unwanted loss
2 In our terminology, extreme adverse event is the most serious category of loss
3 Hindsight is wisdom after the event has happened. It is easy
4 Foresight is wisdom before an event which won't happen if it is foreseen and prevented. It is not easy
5 An accident is the culmination of a process
6 A chain of events is one model of this process
7 The chain of events model is the *Evolutionary Theory* of accident causation
8 An accident is never simply an Act of God
9 An accident is never simply final operator error
10 An accident is a system failure
11 Latent errors accumulate like minor cracks in the system
12 A final event which connects enough minor cracks together to fracture the whole system will trigger an accident
13 A *fatal row* is a line of events and conditions which culminates in an accident
14 In a fatal row, each item in the row is an accident cause
15 Human error is the product of a situation
16 Human error may be forced or unforced
17 Human error is a term too regularly applied to final operators
18 Human error is a term too rarely applied to management hierarchies
19 Root cause analysis is the *Big Bang Theory* of accident causation
20 Accidents end with a big bang more certainly than they start with one
21 The concept of cause is tricky
22 Our definition of accident cause states: *Any failure to take an opportunity to prevent an accident is an accident cause*
23 In general, a system of operation has three objectives and three accompanying risks:
   - Type A objective – conduct of the operation

- Type B objective – avoidance of extreme adverse events
- Type C objective – conformity with regulations, and
- Type A risks – operational risks
- Type B risks – extreme risks
- Type C risks – regulatory risks

25 The cause of an extreme adverse event is the failure to achieve the system objective, Type B. Therefore:

26 The cause of an extreme adverse event is a failure of the system

INTRODUCTION TO PART II: THE LESSONS OF FAILURE

CHAPTERS 5–7

Part II of our book looks for the lessons which we can learn from past failures.

Chapter 5 reviews eleven major accidents. They illustrate different forms of system failure.

Chapter 6 applies hindsight to ten more accidents. It identifies deficiencies in our accident prevention methodology. It shows how and where the systems failed. It shows how and where the decision-making process failed.

Chapter 7 reviews the investigations of two easily avoidable accidents, the first the responsibility of the US Air Force, and the second the responsibility of the Royal Air Force. For different reasons, neither investigation achieved the purpose of accurately identifying the underlying system failures. The RAF investigation did not even identify the cause of the accident. Both investigations therefore failed in their most important purpose, that of eliminating the accident cause.

The message of these three chapters of accident analysis is dispiriting. Those to whom we entrust our lives are consistent, often culpable, in their failure to learn the lessons of accidents. And it is we, not they, who bear the risks of their failure.

CHAPTER 5

A CHAPTER OF ACCIDENTS

WHERE ARE WE GOING?

In this chapter, we look first at risk transfer, the fact that the level of risk to which we are exposed is often decided by others. We review eleven accidents. Each illustrates an important accident cause. You will remember that important cause does not mean either sole cause or root cause; we have bluntly rejected both concepts. The identification of important causes enables us to incorporate appropriate safeguards against them. In other words we can learn from hindsight. It is not difficult. In fact it is easy. Unfortunately, some find it that it conflicts with their interests. Don't fly on their aircraft, sail on their ships, or vote for them.

RISK TRANSFER – THE US MAIL SERVICE

Risk transfer is exactly what it says. Party A decides the level of risk. Party B carries it.

The American Jerry Lederer was a tireless servant of flight safety from the 1920s to the 1990s. In those seventy years aviation went from stick and string in the trees to supersonic in the stratosphere. He fully earned his nickname Mr Aviation Safety. In his book *Aviation Safety Perspectives: Hindsight, Insight and Foresight,* he tells a tale of the US Mail Service. The US Mail was founded in 1918. The aircraft were De Havilland 4 biplanes, fragile contraptions with unreliable engines. They flew at heights that put them in the worst of the weather. (Take it from me, the US has some nasty flying weather.) Bonfires were lit as an aid to navigation. And bonfires were the fatal end of many of the aircraft. In the 1920s Jerry conducted a model experiment to determine why. He ran an empty aircraft down a concrete ramp into a concrete wall and filmed the event in slow motion. The fuel tanks were in the front of the aircraft. The film showed that fuel from the ruptured tanks flowed forward on to the hot exhausts and ignited.

Initially it was up to the local field manager of the Post Office Department to decide whether the weather was fit for flying. Every flight had the urgency of the ride of Paul Revere; the pilots, with First World War aviation heroics fresh in their minds, pressed on. Thirty-one of the first forty were killed in action. In the action not of war, but of delivering the post – can you believe it? The (surviving) pilots rebelled. The outcome was a new procedure: 'if the local field manager ordered the pilots to take off in spite of their better judgement, he was first to ride (in the aircraft) for a circuit of the field'. The ultimate solution to risk transfer. 'Off you go, old boy' gave way to something much more thoughtful.[44] This solution to risk transfer in the US Mail Service will appeal greatly to all who have ever worked at the sharp end. In fact it will appeal to absolutely everyone outside the privileged hierarchies who set the risks for everybody except themselves. We return to them in Chapter 14, The Enemies Within.

RISK TRANSFER – THE FORD PINTO

The (US) Center for Auto Safety gives a particularly clear example of contemporary commercial risk transfer (*emphasis* added):

> In 1977, Mark Dowie of *Mother Jones* Magazine, using documents in the Center files, published an article reporting the dangers of the fuel tank design, and cited internal Ford Motor Company documents that proved that Ford knew of the weakness in the fuel tank before the vehicle was placed on the market but that *a cost/benefit study was done which suggested that it would be 'cheaper' for Ford to pay liability for burn deaths and injuries rather than modify the fuel tank to prevent the*

44 Chicken to Pig: I am willing to participate in egg and bacon. Pig to Chicken: I am unwilling to be involved.

> *fires in the first place.* Dowie showed that Ford owned a patent on a better designed gas tank at that time, but that cost and styling considerations ruled out any changes in the gas tank design of the Pinto.
>
> Closely following the publication of the *Mother Jones* article, a jury in Orange County, Calif., awarded Richard Grimshaw $125 million in punitive damages for injuries he sustained while a passenger in a 1971 Pinto which was struck by another car at an impact speed of 28 mph and burst into flames. Although the award was eventually reduced to $3.5 million by the trial judge, the jury's reason for the figure of $125 million was that Ford Motor Company had marketed the Pinto with full knowledge that injuries such as Grimshaw's were inevitable in the Pinto and therefore the punitive damages should be more than Ford had made in profit on the Pinto since its introduction, which was $124 million.'

In this case the reason for the risk transfer was dead simple. A few dead was the simple solution. It was cheaper than eliminating the cause. The jury set out to reverse that calculation.

In fact the use of cost-benefit calculation is near universal in risk management. It is a perfectly proper tool of decision-making. How else can one decide the allocation of limited resources? However, and most certainly, it is improper to give precedence to cost-benefits over duty to care. The decision was a horrific failing in the value system of the Ford hierarchy; the imperative of human safety was subordinated to the imperative of profit. This case of risk transfer is particularly transparent. Indisputably it was knowing and deliberate. There are other equally shaming cases. However, often the number and probability of deaths consequent on cost saving cannot be accurately or reliably calculated; in such cases it is not easy to prove that executives wittingly preferred profits to lives.

This leaves us with grave difficulties in managing the problem of risk transfer. Somehow we must reflect back on politicians and senior executives the risks which they so happily transfer to us. We need a big fat magnifying mirror. And a trebuchet.[45] Let's work on it.

RISK TRANSFER – THE ARMED FORCES

Risk transfer is the *modus operandi* of the armed forces. Those who sign on do so in the full knowledge that they may have their head shot off in the course of a stupid and avoidable war, or whilst following a stupid or pointless order. I know of no studies that compare the number of deaths in well-considered actions with the number in ill-considered actions. Such a study might have a salutary effect on recruiting. However, many who have served in our armed forces, myself included, have personal experience of unnecessary exposure to danger. If you read the literature of the Second World War carefully you will find that it abounds with accounts of crass stupidity. I know more than one RAF pilot with a distinguished wartime career who regarded the RAF hierarchy with contempt. Too many of their friends were sent unnecessarily to their deaths.

It still goes on. A friend who served in 3 Para in the 1960s tells of the introduction of a new parachute. Many paras regarded it as unsafe. Three refused to jump with it. They were threatened with court martial. One continued to refuse. He was court martialled and sent to Colchester (the notoriously hard military prison). Two jumped; one was killed by the failure of his parachute. The sooner that BIFSS-type military decisions are exposed to public view and to the courts, the better for all, except for stupid officers.[46] And the sooner that military investigations are subordinated to civil oversight, the sooner we will have confidence in their findings. In the armed forces the transfer of unwise risk may equate to a death penalty.

45 Trebuchet. Great big Roman catapult for knocking down walls and chucking things at those behind them.

46 BIFSS – Because I Say So

RISK TRANSFER WILL NOT GO AWAY

Unfortunately risk transfer is a reality of life. General Haig did not lead the advance on the Somme. John Reid did not personally lead our troops into Helmand province to oversee his pious hope of zero casualties; incredible naivety, incredible ignorance, the Afghans have specialized in killing their occupiers for centuries. Tony Blair did not lead patrols into action in Basra. Or anywhere.[47] The Chairman of the National Coal Board did not send his children to school in Aberfan. The Chairman of Union Carbide did not reside in Bhopal. Those who despatched the spaceship *Challenger* were not on board.

In the modern world of large organisations, risk transfer is everywhere. Sadly it is difficult to put the Prime Minister and the defence minister into the lead thin-skinned Land Rover, or to strap an airline chairman to the nose of every aircraft, or to tie the P. & O. Chairman to the bow-doors on every cross-channel ferry. Pity, you say. And you are right. Sadly, injustices are inherent in life. No use complaining. Better to give risk transfer our very careful attention when we formulate our risk management strategy.

THE *TITANIC* – COMPLACENCY

The *Titanic* was a brand new ship. She was supposedly built to the highest standards and incorporated the best of contemporary safety features. She had watertight compartments and the new wireless telegraph equipment. The perception that she was unsinkable was a double agent behind much complacency, including the cost-saving decision to supply only 1178 lifeboat places for the 2224 persons on board. On 10 April 1912 when the passengers arrived for her maiden voyage, workmen were still putting in finishing touches. Her skipper, Captain Smith, was highly experienced. This was to be his last voyage before retirement after forty years at sea. Exceptionally good weather allowed him to sail at high speed; there are suspicions that he may have wished to complete a particularly fast crossing. He continued at

47 To be fair, at the 2005 New Labour Conference his government was a victim of terrorist action, when a verbal charge of 'nonsense' was hurled at them by an 82-year-old.

high speed into a known ice field. The *Titanic* received twenty-one ice warnings in the four days, 11–14 April. The chairman of the White Star line was on board. He suggested to a passenger that the ship might speed up to get clear of the ice.

The Leyland Liner *Californian* was a few miles ahead. She had stopped for the night because of the heavy ice in the area. Her Captain ordered his radio operator to warn the *Titanic* of the ice. This he did. The *Titanic* radio operator was busy with revenue-generating personal messages for passengers and told him to shut up. The *Californian* radio operator, doubtless wondering at this ungrateful response to his safety message, did shut up. He also turned his radio off and went to bed. For that reason the *Titanic*'s distress message went unheard by the ship best positioned to carry out a successful rescue. And where the *Californian* heaved to, the *Titanic* sped on.

Captain Smith had also gone to bed. He left the Sixth Officer, James Moody, in charge on the bridge. At 2340 hr Moody received an iceberg warning from the lookout. He took immediate action and managed to avoid a head-on collision. The iceberg cut a 300-foot gash in the starboard side of the *Titanic*. Modern studies indicate that the damage was exacerbated by the poor quality metal of the rivets.

Unlike some other contemporary ships, the *Titanic* had no longitudinal bulkhead. It had fifteen transverse bulkheads, at least five of which were breached. Worse, the bulkheads were only 10 feet high, so as the ship settled in the water all would progressively flood. One of the ship's designers was on board and he reported to the Captain that the ship would sink in two hours or less. The Captain did not give the order to load the lifeboats for another 30 minutes.

Cause of the accident? A very long list. However, we can see that a thick thread of complacency runs throughout the event. Complacency in design, in construction, in the number of lifeboats and in the response to ice warnings. The Captain's complacency in maintaining a high speed despite the ice warnings was undoubtedly a key factor. Complacency kills. This avoidable tragedy cost 1513 human lives.

R.101 – IGNORANCE

The R.101 was the world's biggest airship. It was innovative in construction. Instead of ribs like bicycle wheels, an outer rim braced by spokes, the ribs of the R.101 were intended to be strong enough to be self-supporting without braces. They were therefore thicker and heavier. The engines were diesels, not only twice as heavy per horsepower as petrol engines, but also necessitating a stronger and therefore heavier structure to carry them. Originally planned to weigh 90 tons without hydrogen or fuel, R.101 finally weighed in at 113 tons. Airships obtain their lifting qualities by filling large internal bags with a lighter-than-air gas. The R.101 contained a vast volume of hydrogen, which is highly inflammable.

In 1924 Britain was still the centre of its far-flung empire. Proponents of airships advocated them as a means wherewith the mother country could achieve rapid and close communication with its distant territories. Two competing projects were commissioned; the R.100 was managed privately; the R.101 was managed by the Air Ministry. Guess what? There were all kinds of failings in the production of the R.101, including the convenient sidelining of the Chief Inspector's warning on its structural inadequacies. Lord Thomson, Secretary of State for Air, staked his reputation on its success. He put his powerful personality into driving the project energetically. Ignoring all warnings that the airship was not ready, he insisted on adherence to his planned departure date. The performance of the hastily completed airship had not been established by a careful test programme. In fact in its final form it had had one test flight, and at reduced power. It had never flown in bad weather. It was relying on hope and hydrogen to keep it in the air.

On 4 October 1930 the largest airship in the world took off from Cardington on its maiden voyage. On to this overweight and patched-up product Lord Thomson had brought a mound of baggage, plus heavy carpets and provisions for a formal dinner in Egypt. Seven-and-a-half hours later, flying at night and in poor weather, it descended and crashed at Beauvais. A massive hydrogen fire immediately consumed the wreck. Of the 54 on board, 48 died; Lord Thomson was amongst them.

Cause of the accident? Just about everything. Over-ambitious concept. Innovative design. High level pressure to achieve declared target date. Grossly inadequate testing. Repeated rejection of warnings. Above all, the ignorance of the Secretary of State for Air. Although ignorant of matters technical, he took the technical decision on the airship's airworthiness for the flight. Ignorance kills. Forty-eight died.

THE DE HAVILLAND COMET – INNOVATION

Take yourself back to 1944, to the last years of the Second World War. Picture the most famous large aircraft of the times, the Avro Lancaster and the B–17. Pause if you wish to review the name the Americans gave to the B–17; the Flying Fortress. It bristled with guns. To little avail: it suffered horrific losses. Hundreds were shot out of the skies by fast and manoeuvrable German fighters. Luftwaffe pilots called the process *Wildschweinjagd*, Wild Pig Shoot. The Luftwaffe remained highly competent until it finally ran out of fuel, pilots and aeroplanes. In the long list of jobs you didn't want in that war, close to the top would be flying daylight missions in a large sitting duck – by whatever name. The Lancaster had four piston engines driving propellers, its wings were straight; maximum speed in level flight was just over 300 mph.

In 1945 great British designers proposed an aircraft powered by the new jet engines and with innovative swept wings. Top speed would be 500 mph. The Comet made its maiden flight in 1949, the first of a new generation of passenger aircraft. US aviation was left nearly a decade behind. By 1952 Pan American was ordering Comets.

The sleek Comet was a paradigm shift in design and construction from the lumbering Lancaster. It was a step into the unknown. Some of the hazards of pioneering soon made an appearance. In October 1952 a Comet failed to get airborne and crashed taking off from Rome. The captain was blamed, so that accident caused no problems to the aircraft's reputation. Then a Comet failed to get airborne and crashed taking-off from Karachi. Investigation showed that over-rotation, that is raising the nose too high at take-off, destroyed the lifting capability of the swept wing. It also showed that the instruments provided the pilot with very poor indication of his angle of rotation. Modifications were made and the problem was alleviated.

Then in the twelve months, from May 1953 to April 1954, three Comets fell out of the sky. The loss of the first, near Calcutta, was attributed to exceptional turbulence in a tropical storm. The second and third were lost after departing from Rome. After the third accident the Comets were grounded and a major and exemplary investigation was undertaken. A structural failure of the wing or fuselage was suspected. A complete Comet was placed in a water-tank at Farnborough and it was subjected to the equivalent of thousands of hours of flying. After more than six months and the equivalent of only 9000 simulated flying hours the fuselage failed. It had proved to be much less resistant to fatigue (repeated cycles of light loads) than had been calculated. At the time the major threat was thought to be structural overload, not metal fatigue. The cabin windows were not circular or oval, as was usual, but rectangular with rounded corners. It was from one of these corners that the crack started.

The detective work that identified the cause was masterly. The only lecture which I can remember from the hundreds of my three undergraduate years was given by a visiting lecturer, a Mr Ripley of the (then) Accident Investigation Branch. He told how the distinctive scalloped edge outline of an Indian anna coin on the aircraft tail had been an important breakthrough. It showed that the tail was still in place when the fuselage failed and the unfortunate passengers and their possessions were hurled to their fates.

The two sources of accident, over-rotation at take-off and metal fatigue as a cause of early structural failure, were to some degree the costs of venturing into the unknown. Whether they were reasonably foreseeable with the knowledge of the times is a matter for historians. What we can say with certainty is that the Comet was innovative. At the time the norm for a large aircraft was a tail-wheel; you lower the nose to take off. The Comet had a nose-wheel; you raise the nose to take off. Its wings were swept; this introduced different lift characteristics. It was pressurised. It flew at new heights. It had unusual windows.

Venturing into the unknown is the essence of progress. No creativity, no innovation: no progress. But the exciting advances of explorers should not too far outpace the dull plod of risk minders.

THE TRIDENT – DEATH BY DESIGN

The De Havilland (DH) Trident would be another nearly great British aircraft. Its sharply swept wing gave it a high speed performance well in advance of its times. However (built-in-but) the greater the sweep of a wing, the worse its low speed performance. It stalls at a much higher speed that a straight wing. When a wing stalls, the airflow, instead of sticking to the wing and creating lift, bubbles away upwards without imparting useful lift. Designers have two answers. The first is a small leading-edge flap which projects forwards and usually downwards from the wing. This often provides a slot in front of the leading edge. The effect is to stick the air more firmly to the wing and thus to increase its lift at lower speeds. The second is a larger trailing edge flap which hangs down behind the wing. Small angles of trailing-edge flap increase the curvature of the wing, thereby improving lift at lower speeds. Large angles improve drag, they slow the aircraft down for landing. Some forms of leading-edge flap pop out automatically as the stall approaches. They were used in early aviation. You will still see them on Tiger Moths, helpfully warning the pilot that he is about to stall. They were also fitted to the Messerschmidt 109 (Me 109), helpfully informing the pursuing Spitfire that his target was running out of lift. The Trident's leading-edge flaps, known as droops, did not operate automatically.

The Trident had a T-tail; that means that the tail-plane and elevator controls were mounted on top of the tailfin. Further, the Trident's engines were at the back of the aircraft. These design features gave the aircraft nasty stall characteristics. In most aeroplanes the nose drops when it stalls; this assists the stall recovery. It gives the doziest pilot a wake-up call and a useful reminder of what he should be doing. In contrast, T-tail aircraft tend to pitch up in the stall;

recovery action requires a much bigger movement. It is also much more urgent. If the stall is allowed to deepen, the turbulent bubbling airflow will envelop not only the elevators but also the engines; the effective operation of both is essential to rapid recovery. The aircraft is then doomed. Exactly that happened to a highly competent test crew on an early Trident test flight.

Normal after take-off drill in the Trident was to establish a safe rate of climb, raise the undercarriage, climb another few hundred feet, raise the trailing-edge flap, slightly reduce engine power, climb to 3000 feet, accelerate to 225 knots – the minimum height and speed for droop retraction – and retract the droops.

Trident G-ARPI took off from London Heathrow on 18 June 1972. It was a typically rotten British Sunday, wet and windy. The BEA crew-room was an unhappy scene of acrimonious dispute about strike action. The captain was known to be brusque and explosive. What was not known was that he had a serious heart condition. The first officer sitting next to him was very inexperienced, and probably apprehensive and nervous. The Trident took off and quickly disappeared into cloud. It was badly buffeted by the turbulence. The autopilot was engaged. It made a very poor job of holding the required speed, allowing it to drop 20 knots below target (far too much). At 1770 feet and 162 knots, the droops were selected in, 1230 feet and 63 knots below the minima. There was a brief opportunity to rescue the situation, but in the confusion of a dysfunctional cockpit the opportunity was not grasped. Who selected the droops up, and why, will never be known.

De Havillands liked neatness in its cockpits. In the Vampire, its early jet fighter, DH grouped the airbrakes, flaps and undercarriage levers neatly together. After landing and before taxiing you raise the flaps, to save them from damage by debris blown up by the jetwash. Guess what? The Vampire proved to be difficult to taxi with the wheels up. When De Havilland placed the levers for trailing-edge and leading-edge flaps side by side on the Trident, as they did, then it was inevitable that somebody sometime would operate the wrong one at the right time, or the right one at the wrong time. It had happened and nearly happened in the past, but quick work had averted disaster. It was a known hazard. BEA captains had long sought a suitable safety device. Unforgivably BEA management had refused to act. On 18 June 1972 this well-known hazard ran through the system's defences and caused the deaths of all 118 occupants of Trident PI. Death by design and by dreadful management decision.

After the accident BEA introduced a speed baulk to prevent the retraction of the droops below their minimum speed. Tombstone Imperative.

THE DC–10 CARGO DOOR – DENIAL

In 1973 I was in the International Air Transport Association (IATA) offices in Montreal. IATA is the trade association of the world's airlines. The Operations and Safety Officer was a friend of mine. Around his small office were boards, each representing an aircraft type. Each board displayed a number of aircraft silhouettes, one for each that had crashed. The largest number, if I remember rightly, was of French Caravelles. The officer pointed to the DC–10 board. 'That's the one which we are worried about.' He explained that his worry related to the engineering of the aircraft, and in particular to the close grouping of vital parts. Damage to a small area would damage many systems.

At 1130 hr on 3 March 1974 Turkish Airlines DC–10 TC-JAV took off from Paris Orly for London. At 1140 hr and at 11,500 feet the rear cargo door blew off the aeroplane. The cargo compartment depressurised. Above it, the passenger cabin was pressurised. The floor of the passenger cabin burst downwards. The lines and cables to the controls in the tail – the elevator, stabilizer and rudder – were fitted directly beneath. They were destroyed. The aircraft nose dropped, the speed increased and 77 seconds later the aircraft speared into the ground at 430 knots. It disintegrated in the massive impact. All 346 on board were killed.

Two years earlier, on 11 June 1972, a cargo door had blown off an American Airlines DC–10 when it was climbing out of Detroit. Very luckily the collapse of the cabin floor left enough controls intact for the aircraft to remain just flyable. The pilots performed brilliantly and the aircraft was landed safely, if not stylishly. When a potentially disastrous defect is identified, the normal procedure is that the aviation authority mandates urgent rectification. In the US the FAA issues an Airworthiness Directive (AD). Aircraft manufacturers greatly dislike ADs not only because they are costly, but also because they imply that the aircraft is not airworthy, aka dangerous. There followed a disgraceful two years of shenanigans by President Nixon's administration, the FAA and McDonnell Douglas. They strongly resisted and finally outmanoeuvred all the efforts of the National Transportation Safety Board and others to obtain an AD and an adequate fix to the problem. McDonnell Douglas had received many warnings about the danger posed by its cargo door design, including one from a subcontractor in 1970. McDonnell Douglas and the FAA acted as if denying the danger.

The Gods cannot be denied. They can be provoked. On 11 June 1972 TC-JAV fell out of the sky. This tragedy had been presaged. It could and should have been easily avoided.

THE *CHALLENGER* SPACECRAFT – PRESSONITIS

Pressonitis is an RAF term for taking a stupid risk, for pressing on regardless of better judgement. It means that if you kill yourself, hindsight will call you a clot.

In certain circumstances of great necessity, notably war, pressing on regardless can be a winning move. The main reason for its success is its total unexpectedness to the enemy. The Persians unexpectedly shinned up the Acropolis at night and caught the Greeks napping. In 1958 the SAS threw away 100 years of military good practice and shinned up Djebel Akhdar in Oman at night and caught the insurgents napping. But in our pursuit of safety we are not fighting dozy humanity, we are fighting the Gods. And the Gods never nap. Therefore in the pursuit of safety it is a rotten idea to press on regardless.

On 28 January 1986 the *Challenger* spacecraft with seven astronauts on board blasted off from the Kennedy Space Center. Just 58 seconds after ignition of the rocket booster a glow was seen from the right booster. This glow was a flame escaping from a seal on the rocket booster. With an exhaust temperature of 5700°F (4000°C) and an exhaust speed of 4000 mph, the booster turned into a mighty cutting torch. The escaping flame cut through the lower strut securing the booster to the spacecraft. The booster whirled round, bursting the external tank. A fireball erupted. At supersonic speed the *Challenger* swung sideways and immediately disintegrated.

Expert engineers at Morton Thiokol, the manufacturers of the seal, had opposed the launch. The material of the seal was known to become brittle at low temperatures. The temperatures at Kennedy were low. The engineers proposed a minimum temperature of 53°F (12°C). This would have postponed the launch. NASA was bitterly opposed to postponement and pressured Morton Thiokol into agreement. The Morton solid rocket manager refused to sign the release, so his vice-president signed instead. During the countdown before blast-off the temperature of the left booster was 33°F (1°C) and the right booster was even colder, 19°F (–6°C). The seal was outside its recommended operating limits. Unsurprisingly it failed.

The sociologist Diane Vaughan wrote a detailed and highly praised book on the accident. She attributed it to 'the normalisation of deviance'. What on earth does this sociologist-speak mean? It means that deviation from safe procedures had become the normal practice, an example of what we have called shifting sands. She is not wrong but in my view she is not right either. Her ponderous expression is an accurate description of the *context* of the decision to launch. It does not address the decision itself, a decision which was insupportable.

The decision completely subordinated the imperatives of risk management to other imperatives. In this case, to maintaining the credibility of the space programme. The managers used their managerial power to bulldoze away the objections of their specialist risk managers, the expert engineers. Amazingly the final decision to launch was taken, not by the aircraft commander, but by ground managers. Unforgivably the decision and its circumstances were not even submitted to the commander for his approval. Other than in the military in wartime, the acceptance by its captain of the aircraft's safety for flight is always required. Even in wartime he would at least be fully informed of its state. It is not within the captain's capability to understand all the technical *minutae* of his aircraft. His job is to consider whether he can, and how he will, deal with a technical failure. Had the matter of the operating limits of the seal been referred to Dick Scobee, the *Challenger*'s commander, I believe that his reply would have been short and possibly rude. And he and his brave colleagues would have been alive today. Risk transfer: NASA had forgotten the lesson of the US Mail. Pressonitis kills. In this case it was the pressonitis of the management. But it was the crew who were killed.

KOREAN AIR 747, STANSTED – CULTURAL DEFERENCE

On the dark night of 22 December 1999 a Korean Air 747 cargo aircraft took off from Stansted. Four minutes later, it struck the ground in a steep descending turn, killing the four persons on board.

The inbound crew had reported a problem with the bank indications on the Captain's attitude indicator (ADI). The Korean ground engineer was not well qualified in electronics. With the assistance of a qualified engineer from the ground handling company, he had investigated the attitude indicator and found and corrected a fault.

The captain did not encourage input from his crew. As the aircraft taxied out, he sharply rebuked the first officer for his careful, allegedly incorrect, radio procedure. Worse, it was the captain who was ignorant of the procedure; the first officer had been correct. In any case the captain was seriously at fault for discouraging care and caution in a young pilot. The martinet personality of the captain strongly reinforced the customary Korean cultural deference to rank. With a heavy atmosphere on the flight deck, the aircraft departed.

A night take-off in poor weather is the worst possible way of testing the success of a repair to an ADI. It is the most important instrument on the aeroplane. It tells the pilots which way the aircraft is pointing. The serviceability of his ADI should have been uppermost in the captain's mind. He had a range of safety opportunities open to him. He should have briefed his crew of the dangers of the situation. He should have asked the first officer to carry out the take-off. He should have instructed his crew to monitor his ADI with extreme care, cross-checking it against the other two attitude indicators, etc. He did none of these. He just took off. What the crew did not know was that the serious fault with his bank indication was still present. It was a different fault which had been corrected. The captain was preoccupied with adhering correctly to the departure procedure. It was a simple procedure. He did not realise that his ADI was effectively frozen in bank. So he rolled into his procedural turn. And went on rolling. The nose of the aircraft dropped and the 747 entered its fatal dive.

The first instrument warning sounded 25 seconds after the aircraft took off. Then 19 seconds later the flight engineer reported, 'Bank not working'. A further four seconds later, he repeated, 'Bank, bank'. Finally, 15 seconds later, with the aircraft hurtling towards the ground and with the increasing wind noise clearly audible on the cockpit voice recorder, the engineer again said, 'Bank'. 'His last remark before impact ... appeared to be said with resignation'. One and a half seconds later they were all dead. At no time did the captain respond to the flight engineer's calls. At no time did the first officer intercede, although at the very last moment, far too late, he may have made a control input. The crew's performance was unbelievably bad. Judging by his performance that night, the captain had no place in a complex multi-crew aircraft. His colleagues failed absolutely to follow the simple procedures which would have rescued the situation, even with such a captain. It appears that both the Korean Air operational procedures and crew training were deficient. The airline was seriously at fault.

The failure of the first officer and flight engineer to intervene with forceful words or actions is primarily attributable to their cultural deference to rank. As their aircraft plunged earthwards, this deference was a greater imperative than saving their lives. Unbelievable. In a complex multi-crew aircraft, it is a primary duty of the captain to weld his crew into a team. The safety of the aircraft is obtained only when each team member is both competent and committed in his role. It is essential that each monitors the actions of his colleagues, thereby to ensure that they are safely performed. So we take a mental note never to board the aircraft of any country in which deference and obsequiousness is the cultural norm.

And we breathe a sigh of relief that such an accident could never occur in the well-reputed airlines of Europe and America.

Sure? You may wish to locate a source of strong drink before reading on.

TENERIFE 747S COLLISION – AUTHORITY FIGURE

The Canaries is a group of islands off the west coast of North Africa. It is a Spanish possession. In 1977, as was the fashion of the times, it harboured a terrorist group with a weighty title, the Movement for the Independence and Autonomy of the Canaries Archipelago. This group is believed responsible for the terrorist bomb which exploded in a florist's shop at Las Palmas airport on the east coast of Grand Canary Island on Sunday 27 March 1977. No one was killed by the bomb itself, but its unexpected consequences led directly to the world's worst air accident.

Two Boeing 747s, KLM flight KL 8405 and Pan-American Clipper 1746, were inbound to Las Palmas when the airfield was closed after the bomb blast. Both were diverted to Los Rodeos airport on the nearby island of Tenerife. Los Rodeos is a small airport with a single runway and one parallel taxi track. The KLM landed first and the Pan-Am landed 37 minutes later. Both flight crews were inconvenienced by their diversions. They were impatient to get away quickly to avoid serious disruption to their schedules. Soon afterwards Las Palmas reopened and Pan-Am sought an immediate departure. The Pan-Am 747 was too big to pass the KLM 747. KLM was not ready to move. It was not until nearly two-and-a-half hours later that KLM received clearance to taxi. The Pan-Am followed.

By then the Pan-Am crew had been on duty for more than 11 hours and the KLM crew for more than 9 hours. Both crews were tired. Thick fog was forming in patches. Neither crew knew the airfield. The several short links between the runway and the parallel taxi track were potentially confusing. The ground controller spoke with a heavy accent and communications were not easy. The KLM was instructed to backtrack the runway to the end, make a 180° turn and await instructions. The Pan-Am was instructed to follow the KLM but to turn off the runway via one of the links to the parallel taxiway. Pan-Am's request to hold position and remain clear of the runway until after the KLM had departed was refused. There was a thick patch of fog on the runway. The KLM turned at the end of the runway and reported 'We are now at take-off'. The controller replied, 'OK … . stand by for take-off … . I will call you'. At the same time Pan-Am said 'We are still taxiing down the runway'. The two simultaneous transmissions caused some radio interference, but did not make them unintelligible in the KLM cockpit.

The KLM captain was a captain of great authority and prestige. He held the powerful position of Head of the Flight Training Department at KLM. He appeared in KLM publicity material. He had given the first officer his initial qualification check on the 747 only two months earlier. It is likely that the first officer was daunted both by the captain and by his own inexperience.

As soon as he had completed his 180° turn and lined up, the KLM captain started to advance the throttles for take-off. The first officer stopped him, 'Wait a minute, we do not have ATC clearance (permission)'. The captain closed the throttles, responding ungraciously, 'No, I know that, go ahead, ask'. The first officer then took down the ATC clearance; it did not include the vital words 'cleared for take-off'; it did include the phrase 'after take-off'. While the first officer was still reading back the clearance, the captain opened the throttles and commenced his take-off run. Pan-Am was still on the runway, hidden in the thick fog-patch. Pan-Am responded to an ATC call, 'OK, we'll report when we're clear'. The KLM flight engineer said, 'Is he not clear then?' The captain did not understand and he repeated, 'Is he not clear that Pan-American?' The captain responded with an emphatic 'Yes'. The first officer and the flight engineer made no further objections. Thirteen seconds later the KLM captain failed in his frantic effort to fly his 747 over the Pan-Am 747. A total of 583 persons were killed, including all 248 on board the KLM. It was the worst crash in aviation history.

It is clear that both the first officer and the flight engineer on the KLM were concerned by the captain's assumption that the runway was clear. Either could have prevented the accident. They could have spoken up firmly, or could simply have pulled the throttles back. They did neither. They subordinated their personal reservations and professional judgements to the captain's status and to his personal prestige. They deferred, not as did the Koreans, to national culture, but to the authoritarian culture in their cockpit. Same result. An avoidable, unnecessary accident. And at Tenerife appalling cost.

THE DC–10 SLATS – IMPROBABLE FAILURE

On 25 May 1979 an American Airlines DC–10 was taking off from Chicago O'Hare. Just before lift-off the number 1 engine became detached. It tore away a section of the leading-edge on the left wing and destroyed a number of vital electric and hydraulic controls. Unbeknownst to the crew this caused the slats on the left wing (leading-edge flaps, similar to the Trident's droops) to retract. The slats on the right wing were unaffected. Worse, the warning system which would have informed the crew of the slat asymmetry was amongst the several systems rendered inoperative by the damage. Otherwise the loss of the engine would probably have been survivable. Further, the pilots were probably unaware of the nicety that the stalling speed of a wing in this condition was extremely sensitive to small angles of bank. Twenty seconds after take-off the left wing began to stall. Control of the aircraft was lost. It rolled to the left; the nose dropped. It hit the ground 31 seconds after it had taken off. All 271 on board and 2 more on the ground were killed.

McDonnell Douglas considered that the separation of an engine was as mathematically improbable as the structural failure of a wing and therefore felt under no obligation to address it. Analysis by

McDonnell Douglas had indicated that simultaneous engine failure and asymmetric flap retraction might compromise safety margins, but that it was extremely improbable. The manufacturer also felt under no obligation to address that. Apparently they did not reconsider this decision even after manufacturing deficiencies and maintenance damage affecting engine pylons became known. The probability of the simultaneous occurrence of three events: engine failure, slat damage, and at take-off, was deemed to be less than one in a billion. By September 1981 it had happened four times. McDonnell Douglas then fitted a locking mechanism to the slats. The Boeing 747 and the superbly engineered Lockheed Tristar had long had such a device as standard.

There must be grave doubts about McDonnell Douglas decision making. Engine separations had happened in the past, as had damage to engine pylons. Further, it was by no means unknown for fan blades and larger chunks to fly off engines. Engines are at maximum power and under maximum strain on take-off. The vulnerability of the concentration of vital systems in the wing leading-edge should have been self-evident. Most certainly it should have been obvious that slat asymmetry would be potentially fatal during the take-off phase.

The mills of God may grind exceedingly slow but they grind exceedingly fine. If there is one chance of an accident in ten million events, then the probability of an accident on the first occasion of the event is one in ten million. It is very unlikely. What can go wrong, usually goes right. Good, you say. A small yes, but for one reason only – we have got away with it once. Two big nos – we have learned nothing of the danger lying in wait for us and worse, we have taken the first step on the road to complacency. The more often that we get away with it, the more complacent we will become; the more our guard will drop and the higher will become the probability of an accident. If the mills of God ground faster, we would learn faster. But they grind slow, so we learn slow. It is a paradox, but the Gods write the rules for their benefit, not for ours. They encourage us to think that they are not watching. But they are, and they are biding their time. And enjoying the foreplay. One in ten million means that it will probably not happen this time. But happen it will.

The simple commonsense of my friend at IATA who I mentioned in our discussion of the DC–10 cargo door provided a much shrewder assessment of risk than all the high mathematics of one in a billion calculations. Three foot of damage took out a concentration of systems, and that took out the whole aircraft and all its occupants. McDonnell Douglas and the FAA hoped that the Gods would not be watching and waiting. My friend at IATA knew that they were.

CONCLUSIONS

At this stage we draw no conclusions from these eleven cases. In the next chapter we summarize more briefly another ten accidents. We then will have a large enough number to draw general conclusions. We will see what lessons we can learn. There are several, and they are useful.

CHAPTER 6

A CHAPTER OF CONCLUSIONS

AND OUR POINT IS

'And your point is?' is a useful question. It can be deployed to bring a speaker back to the point. Or to a halt. Politicians and other clever dicks make clichéd use of it to interrupt unwelcome opposition, and to infuriate. Let us therefore declare what our point is, before the critic or the unworthy (or the unworthy critic) asks.

Two of the conclusions which we reached in Part I were that:

1 an accident is a failure of the system of operation
2 three of the categories of failures of systems are:

- self-inflicted injuries
- double agents
- shifting sands

In this chapter we will test these two conclusions against each of the eleven accidents reviewed in Chapter 5, and against ten more accidents which we briefly review in this chapter. The results of this analysis, Table 6.1, Nature of System Failure, are presented in easy form in the appendix to this chapter, where all tables will be found.

Next, to establish in which system function the system failure occurred, we create a simple model.

First, we break the system of operation down into four system functions:

- Design

- Direction
- Management
- Final Operator

Second, we determine in which of these four system functions, the system failure occurred.

Results in easy form in Table 6.2, Locus of System Failure.

As you read this chapter's review of ten more accidents, you may wish to test for yourself our two conclusions from Part I and our simple model.

The process of examining these failures of our defences leads to an interesting, and possibly unexpected, finding. At the heart of most accidents is a bad decision.

This draws our attention to decision-making. As a first step we construct a simple model of decision-making in risk management. We have already suggested that the success of a decision rests upon three key components:

- Information
- Intelligence
- Integrity

This model can be likened to a three-legged stool; each leg represents one component of decision. The stool fails with the failure of any one of its three legs. The risk management decision fails with the failure of any one of its three key components. We test this model against our accident sample. Results in easy form in Table 6.3, Nature of Decision Failure.

Tables 1–3 are combined in Table 6.4, Overview of System Failure. For each of our twenty-one accidents Table 6.4 shows:

The nature (self-inflicted injury, etc.) of the system failure

The locus (system function: direction, etc.) of the system failure

The nature (information, intelligence, etc.) of the decision failure.

Our sample of twenty-one accident case studies is large enough to give us confidence that our findings may have some general validity.

Our tests suggest that we are on the right track. We are beginning to see how systems fail. We can draw our initial conclusions on the prevention of accidents.

And that was our point.

SIMPLE MODEL OF THE SYSTEM

We now need to take the concept of a system of operation one step further. This calls for a first analysis of the system's constituent parts. There are numerous possible bases for such analysis. A hierarchic basis yields an organisation chart; a chronological basis yields a flow chart; a cost basis shows where the money goes; etc., etc. Each has its specific uses. None provides a complete analysis.

We need a very simple picture of the functions of the system of operation. Let us subdivide it into four main functions: design, direction, management and final operator.

Design is the design of the fundamental vehicle or vehicles of the operation. Vehicle in the sense of the fundamental machinery on which the operation is based. In transport, the ship or airship or aeroplane. In manufacturing and processing, the plant. In bureaucracy, the system. In computing, the hardware and the software. Remember that we are interested in system failure. The vehicle in which we are interested is the one involved in the failure.

Direction is the highest level of control of the operation. It involves the functions of laying down the organisation's targets and values, of setting the task of the operation, and of allocating its resources.

Management covers all functions which lie between the direction of the operation and the functions of the final operator.

Final Operator function is self-explanatory. It covers all functions entrusted to the final operator.

We want only a general picture. Therefore we are not going to involve ourselves in the careful definition of the dividing lines between the functions. They are not important to our purpose.

Take the *Titanic* as an example.

Design refers to the functionality of the ship and of its equipment, its fitness for purpose; therefore its seaworthiness, its bulkheads, its robustness.

Direction refers to setting the organisation's style (targets and values), to the deployment of the *Titanic* on the North Atlantic (task) and to deciding its equipment, including its complement of lifeboats (allocation of resources).

Management refers to all functions between the directors and the ship's captain.

Final Operator functions are those of the ship's captain, included those which he delegated to his subordinates.

Let us now review ten more accidents.

OCEAN RANGER OIL RIG – UNSINKABLE

When it was built in 1976, the *Ocean Ranger* was the biggest oil rig in the world. It was designed to withstand 100 foot waves. On 14 February 1982 it sank in a storm off Newfoundland. The failure sequence was started by a 50 foot wave. This smashed a small glass porthole in the ballast control room. Ballast control keeps the rig level and at the correct depth; the system was supposedly fail-safe. It was not. The sea entered and disrupted the electric control systems. The rig began to tilt. Seawater began to enter through openings which would normally have been high above the water. No way of closing them had been provided. The personnel were not well-trained; nor did their manuals cover the emergency with which they were faced. Escape drills had never been tested in severe weather; it proved impossible to launch the sophisticated lifeboats safely, or to transfer from them to rescue vessels. All 84 on board were lost.

Ocean Ranger sank because it was considered unsinkable.

THE BHOPAL DISASTER – FAILURE OF SELF-REGULATION

Few disasters, however horrifying, kill many people. The *Challenger* killed seven, the DC–10 killed 273 at Chicago (slat accident) and 346 at Ermenonville (cargo door). Road accidents and medical accidents kill thousands every year, but each event, however reprehensible, involves one or a few deaths. Thankfully, man-made disasters which kill thousands at one time are very rare. In general terms, only dam failures and nuclear power and chemical process accidents have that potential.

On 3 December 1984, in the world's biggest chemical disaster, thousands (estimates vary hugely; Greenpeace puts the figure at 22,000, others at between 2500 and 7000) were killed at Bhopal, India. Many victims were packed in a densely populated shanty town around the Union Carbide Corporation plant. Their manner of death was horrifying, choking on poison gas, banned even in wartime by international convention.

There had been safety warnings at the plant, and there were various problems in the chemical process. The well-informed company or government safety supervision which would have prevented disaster was simply absent. The Indian government may not have had the necessary expertise; the company certainly did. Three out of the four relevant safety systems had been shut down by plant operators.

Union Carbide attributed the accident to sabotage. This explanation is not widely accepted. A Union Carbide vice-president said that 'there were maintenance problems that would not have been tolerated at a plant in the United States'. In the light of this statement, Union Carbide would be hard-pressed to prove that it had exerted itself diligently to prevent accidents at its Indian subsidiary.

The disaster at Bhopal left a nightmare legacy, birth defects which occurred for years afterwards. It was a costly and dreadful failure of self-regulation.

USS *VINCENNES* – DISASTER BY INCOMPETENCE

In the late 1980s the US Navy was the source of grievous concern to airline pilots flying over the Gulf. The deployment of the fleet gave every impression of being an operation badly planned and dangerously well-armed. For example. One dark night sitting at 33,000 feet in my BA Tristar I received a call on emergency frequency from the US

Navy, 'Aircraft flying in the area of the US Fleet, change course immediately'. Thinking that I might be that aircraft, I answered, 'US Fleet, this is Speedbird ***, what is your position?' Reply, 'Negative, Speedbird, US Fleet position is a military secret'. Script by Spike Milligan. So I, and probably a good number of other airliners, rapidly changed course just in case. On another occasion, a colleague low on his approach to Dubai was ordered to overshoot. All of which spells real danger to civil aircraft. A major problem was that the US Navy had no radios which operated on normal airline frequencies. In those days airline security was my business, so on behalf of the British Airline Pilots Association, I offered to buy the radios for them.

On 3 July 1988 the USS *Vincennes* shot down an Iran Air Airbus. The Airbus took off from Bandar Abbas, a joint civil and military airfield. It was transmitting the normal civil code on its transponder. Nonetheless the *Vincennes* mistook the decreasing range of the Airbus for decreasing height, and interpreted it as a fighter diving to attack. The performance of the *Vincennes* was not creditable, perhaps the crew's training was inadequate. The most serious culprits were elsewhere. The US Command should have established good radio contact not only with the airports of the friendly Gulf States, but also with the Iranians, and thus been continuously well-informed of all civil aircraft movements. As a second best they should have been monitoring the Iranian frequency. The Iranians should have had the sense not to have operated fighter aircraft off the same airfield as civil airliners. Damnable incompetence on both sides. And 290 innocent passengers and crew suffered the consequences.

Disaster by incompetence.

PIPER ALPHA OIL RIG – FAILURE OF PROCEDURES

On 6 July 1988 an explosion on the *Piper Alpha* North Sea oil and gas platform killed 165 of the 226 on board. 'Communication failures at shift handover ... together with a breakdown of the (relevant) permit to work system' were cited. As important but more fundamental were the findings of the Public Inquiry chaired by Lord Cullen. His lengthy report condemned the failure of the safety inspectorate of the Department of Energy. Lord Cullen proposed a new structure for the management of safety. His model was widely adopted.

The accident to *Piper Alpha* occurred because there were serious inadequacies in the safety of its operating procedures.

LOCKERBIE – PRIORITIES

In 1988 aviation security was struggling. The task of safeguarding aircraft operation world-wide was immense. Terrorist attacks were relatively rare. The security counter-measures were onerous and disruptive. Their efficacy was open to doubt. Within airlines there was considerable commercial pressure to cut security corners. There were many merry battles with aircraft captains who for obvious reasons were determined to maintain the best possible security standards on their aircraft. On 22 December 1988 the Pan-Am Boeing 747 *Maid of the Skies* was blown apart over Lockerbie by a terrorist bomb. It is widely believed that the bomb was placed in a suitcase which travelled from Malta to Frankfurt. In Frankfurt it was transferred without the regulatory X-ray screening to a Pan-Am container which was then flown on a Pan-Am aircraft to Heathrow; there it was transferred without screening on to the 747 *Maid of the Skies,* Pan-Am Flight 101. Pan-Am alleged that the US Federal Aviation Agency (FAA) had authorised such unscreened baggage transfers in Frankfurt. Neither Pan-Am nor the FAA emerged from the investigation with credit.

It is clear that had the lethal bag been subject to adequate X-ray by proficient staff at Frankfurt (note the two big ifs, adequate and proficient), the bomb would have been detected. Further, the captain of Pan-Am 101 was not informed that bags had been loaded on to his aircraft without the normal security screening. So he had no opportunity to order their offloading.

The economics of airlines are fragile. It is very easy for them to lose money, lots of it. Pan-Am had not been truly profitable in years. Aircraft delays are costly and inconvenient, therefore corners are cut to achieve punctuality. Had costs and convenience not been given priority over security, the bomb on Pan-Am 101 would have been detected. If gaps are left in defences a persistent or a lucky antagonist will sooner or later find them. In the Lockerbie atrocity 271 died.

Economics were only part of Pan-Am's security inadequacies. The airline had chosen to spend much of its US government security subsidy, not on security but on the cosmetic of reassuring the public of its security. Subsequent to Lockerbie Pan-Am went out of business. Some would say deservedly.

HERALD OF FREE ENTERPRISE – DIRECTION

Nearly all accidents involve a degree of mischance, usually the unfortunate coincidence of a number of factors. With this the system of operation fails to cope. In the accident to the *Herald of Free Enterprise*, the mischance was minimal. The system of operation did not cope because it was grossly flawed, simply unfit for purpose.

On 6 March 1987 the *Herald* sailed from Zeebrugge with her bow-doors open. The sea came in. The ship capsized and 188 were drowned in the chilly waters; most were trapped inside.

1 There was no system to ensure that the doors were closed
2 There was no system to report to the bridge that the doors were closed
3 There was no indicator system on the bridge to show that the doors were closed
4 The ferry was routinely overloaded
5 There was no draught indicating system
6 There was no permanent complement of deck officers; crews were collected *ad hoc*

Masters (the ships' captains) had more than once informed the company's directors of all six of these deficiencies. And 'Before this disaster there had been no less than 5 occasions when one of the Company's ships had proceeded to sea with bow or stern doors open.'

When this accident occurred the civil aviation community viewed it with incredulity. We just could not believe that none of the most elementary safety procedures were in place. But they were not.

Mr Justice Sheen who conducted the Formal Investigation was considerably exasperated by the conduct of some the company's directors, both as directors and as witnesses to his court.

> The Board of Directors did not appreciate their responsibility for safe management of their ships ... The directors did not have any proper comprehension of what their duties were ... From top to bottom the body corporate was infected with the disease of sloppiness ... This became particularly evident from the evidence of Mr ... , the Technical Director and of Mr ... , the Operations Director .... the court was singularly unimpressed by both these gentlemen ... staggering complacency ... vacuum at the centre ... shore management took very little notice of what they were told by their Masters ... his statement ... has a hollow ring; an absence of a proper sense of responsibility ... (a director of the company) was a very unsatisfactory witness ... verbose, rambling and at times ... misleading.

The investigation revealed that the ships' captains were under heavy company pressure to accept the company's defective procedures. Nonetheless Mr Justice Sheen found that 'the accident was partly caused or contributed to by the serious negligence of the Master, the Chief Officer and the Assistant Bosun'. He suspended the certificates of the Master and the Ship's Officer for twelve months.

He found that 'the accident was partly caused or contributed to by the fault of Townsend Car Ferries Limited (the owners)'. No, no, no, Mr Justice Sheen. Not partly. The extraordinary unconcern of the company directors with elementary safe practice had set a scene for guaranteed disaster. They imposed a context in which, sooner or later, the occurrence of an accident was inevitable. On 6 March 1987 the unfortunate captain of the day failed to prevent the accident which the directors had arranged; 188 died.

What an example of free enterprise. Mr Justice Sheen should have suspended the company directors. By the neck.

VALUJET – TOO FAR, TOO SOON

On 11 May 1996 a ValuJet DC–9, Flight 592, crashed into the Everglades swamp in Florida. All 110 on board were killed. ValuJet was one of the new generation of cut-price airlines. The cause of the crash was smoke and fire in the aircraft fuselage. The source of the smoke and fire was a consignment of oxygen generators in the hold.

These generators are standard aircraft equipment, used commonly to supply the oxygen masks which drop down in front of passengers if cabin pressure is lost. They generate oxygen by a chemical process during which their surface reaches 500°F (260°C) in normal usage, more in a confined space. Because the handling staff did not understand their dangers, they were packed tightly into the aircraft hold, near to three aircraft tyres, at least two of which were on their wheels and inflated. The handling agents were subcontractors; they employed staff with limited English. The difference between the words 'expended' and 'expired' on the labels attached to the oxygen cylinders was the difference between safe and dangerous; 'expended' means 'discharged' (safe), 'expired' means 'beyond approved life' (unsafe); this subtlety was the difference between life and death for the occupants of ValuJet 592. It was the first of many fine but important details which lay deep in ValuJet's engineering instructions and operating procedures, hidden from all but the experienced expert eye. On 11 May there were no experienced eyes where they were needed.

Once again the FAA was savagely criticised. Its inspectors had indeed reported serious reservations about the ValuJet operation to their Head Office but had elicited no response. The FAA was saddled with two conflicting objectives, promoting the airline industry and enforcing safety. Safety campaigners had long held that its relationship with the airlines was far too cosy.

ValuJet had expanded so fast that its operation had far outpaced its infrastructure. It did not have an organisation adequate to the huge task of assuring the safety of its complex activities. ValuJet was concentrating on responding to the market opportunity; it had lost sight of the importance of avoiding extreme adverse events. Objective

A (conduct of the operation) had overwhelmed objective B (avoidance of extreme adverse events).

AIR FRANCE CONCORD – DENIAL

The Concord is not a big aeroplane. Close to the tyres of the main wheels are the hydraulics and the electrics of the aircraft's control and warning systems; and the engines with their huge high-speed rotating assemblies, flaming combustion chambers and highly flammable fuel at high pressure. All fine when in its right place, but not fine when any of it gets out. A tyre explodes like a bomb, chunks fly like shrapnel. That is what happened to the Air France Concord on take-off at Paris on 25 July 2000. As ever there were a number of mischances. A small piece of metal on the runway damaged a tyre. It exploded, 'punched holes in the wing, leading to a massive fuel leak, a fire and other damage'. The aircraft rapidly became unflyable; it crashed, killing all 109 on board and 4 on the ground.

Concord had suffered tyre problems all its working life. It was a tyre failure that led to the Air France Concord at Paris becoming uncontrollable. This disaster permanently terminated all Concord operations.

An accurate history of all Concord tyre failures which caused aircraft damage is not easily obtained. Because the Concord was an iconic aircraft and a prestige flag carrier, the French and to a lesser extent the British are coy about precise figures. Andrew Weir, author of the dolefully titled *The Tombstone Imperative* quotes 'at least six potentially catastrophic fuel leaks in under 80000 flights'. These were associated with tyre failure. Note and contrast the 6 in 80000 (approximately 1 in 14,000) with the 1 in 1,000,000 or 1,000,000,000 probabilities normally associated with catastrophic failure in aviation.

Some BA Concord pilots allege that the accident was the result of a chain of Air France deficiencies: carriage of extra baggage and fuel, possible tailwind, failure to replace a spacer on a wheel which consequently overheated and ignited spilled fuel, etc. Their oblique message is that BA would never have had such an accident. May be. But may be not. The aircraft's history of tyre bursts and fuel escapes was a history of near-misses. The accident was waiting to happen.

When it did happen, engineers proposed a more stringent tyre regime and the armouring of the fuel tanks with Kevlar. So why had not these proposals been made before? Perhaps because they would have been most unwelcome. Adding weight to any aircraft reduces its payload; it is always fiercely resisted. The icon had always been close to unviable, its fuel capacity barely adequate, its payload minimal. The extra weight of the Kevlar would have reduced them further. For that reason the probability of disaster was denied.

The aircraft's supersonic performance made it an icon. It dictated the lightness of its structure. And it made it vulnerable to a simple tyre burst.

THE NHS: QMC – CLUELESS

On 3 January 2001 a patient attended the Queen's Medical Centre, a large hospital in Nottingham, for an intrathecal injection (IT, in the spine). He was injected with a drug suitable only for injecting in a vein (IV). Sad to relate, the patient contributed to his own demise. Known to be unreliable, he arrived in the afternoon for a morning appointment. Instead of his fitting into a smooth (we hope) hospital routine, the staff made *ad hoc* arrangements in an endeavour to accommodate him. He died the next day.

We will look at this accident in more detail in Chapter 8. There we identify no less than twelve broad categories of opportunity to have averted the accident. For the moment we note only the gravest failing. The IV syringe fitted the IT spinal needle.

If it is physically possible to connect the wrong syringe to the spinal needle, then sooner or later someone will do it. Basic, elementary, Murphy. A mechanical safeguard, equally basic and elementary, would have prevented the accident: a dedicated IT spinal needle to which nothing but an IT syringe could be fitted. Such safeguards have been incorporated in aviation procedures for over fifty years. Yet not adopted by the NHS in the twenty-first century. Incredible.

Not only did the QMC patient die, but worse, he was killed by a known danger. And, even worse, identical accidents are still occurring in hospitals to this day. But the official verdict? Send the senior doctor to prison. True, he did not do very well. But, as we will see in Chapter 8, those responsible for the safety of the system did much worse. The senior managers responsible for procedures, the senior consultants cognisant of the dangers, the senior civil servants and the ministers who set the NHS safety standards and priorities didn't do much good, and they certainly and sadly didn't do time. The sooner they find themselves in the firing-line, the better. Let us hope that someone sues them to their back teeth very soon.

THE TWIN TOWERS, 9/11 – INERTIA

Some Middle Eastern men turned up at a US flying school. They wanted to learn to fly, but were not particularly interested in learning to land. Not a clue for a bright man, a clue for any idiot. Flying instructors are not idiots so they told the FBI. And somewhere in the FBI there was enough bureaucracy or enough idiots for this vital intelligence to go to waste.

In security all risk assessments start with four questions:

1 Can they do it?
2 Do they want to do it?
3 Does it fit with their known *modus operandi*?
4 Have they opportunity and reason to do it soon?

If you get four yeses, either wake up with a bang, or be woken by a bang. The intelligence from the flying school was the fourth yes. The US authorities did not wake up until 11 September 2001. Then they did.

Let us not make excuses, but responsibility for security is a poisoned chalice. No one can ever guarantee total security. A determined adversary, let alone a suicide attacker, will always get through somewhere. To respond to every possible threat will eat more resources than any country has available. It is pointless for security chiefs continually to request resources from ministers: they won't get them; but they will rapidly become unpopular and, very probably, will be eased into early retirement. Be that as it may, the prime duty of the US intelligence bureaucracy is to prevent extreme adverse events. It could have prevented 9/11. It failed. Positive action drowned in the sticky sludge of bureaucratic inertia.

CONCLUSIONS: THE NATURE OF SYSTEM FAILURE

1 *Self-inflicted injury* underlies the system failure in twenty of the twenty-one accidents

2 Self-inflicted injuries are cases in which hubris, complacency, insufficient thinking, ignorance, or plain stupidity, leave serious gaps in the system defences, open invitations to the Gods to disrupt our activities. By and large the resulting accidents are unforgivable

CONCLUSIONS: THE LOCUS OF SYSTEM FAILURE

1 To pinpoint the *locus of the system failures*, we sub-divided the system into design, direction, management, and final operator. These functions closely relate to chronological phases

2 The earlier the phase of a failure, the more the opportunities to rectify the initial failure There may be no opportunity for rectification of a serious error by a final operator, as in the accident at Tenerife. However there is nearly always opportunity *before* the event for good procedures to prevent the fatal error. Had KLM crew training emphasised that aircraft safety must be given priority over deference to authority, there would have been no accident at Tenerife. Thus, both the failure to rectify an early system error and the failure to prevent irrecoverable error by a final operator are system deficiencies

3 Twenty of the twenty-one accidents involved failures in more than one function or phase. This is powerful evidence for the concept of an accident as the complete failure of system defences

4 Consider the widely cited finding that 70–80% of accidents are attributable to final operator. Compare that with our findings:
   - 19% attributable to final operator error
   - 57% attributable to a design failure
   - 95% attributable to a failure of direction and/or management, if we exclude the Comet. There is a good argument that the Comet also was a failure of direction, which would make 100%.

5 Make that comparison and tell the establishment where to put its figures of final operator error

6 And tell the courts what to say to the complacent, the profit seekers, the directors and the managers for whom your safety is secondary to their comfort. Mind your fingers. Clang

CONCLUSIONS: THE NATURE OF DECISION FAILURE

1 We likened a risk management decision to a stool, resting upon legs of information, intelligence, and integrity

2 We find that there were two failures of information, eight of intelligence and eleven of integrity

Our analysis owes as much to common-sense as to scientific rigour. Our sample is too small for statistical analysis. Nonetheless we can draw provisional, useful, conclusions.

GENERAL CONCLUSIONS

A review of our twenty-one case studies reveals the following:

1 Self-inflicted injury is the gravest weakness in our defences. It occurred in every case with possible exception of the Comet. We leave obvious gaps in our defences which positively invite the Gods to put the boot in

2 Complacency (it will not happen) is the most common form of self-inflicted injury, occurring in 19 out of 21 cases

3 The direction function is the most common locus of system failure, 18 out of 21

4 All accidents originating in failures in the design function, with the possible exception of the Comet, could have been prevented by responsible action in the direction or management functions
5 All accidents involving failure of the final operator could have been prevented by responsible earlier action in the direction or management functions
6 The failure of integrity is the most common form of decision failure, 11 out of 21 cases
7 All failures of integrity occurred in the direction and/or management functions
8 The failure of intelligence is the second most common form of decision failure in our case studies, 8 out of 21
9 The decision failure was by the final operator in 4 cases (*Titanic*, Tenerife, Korean 747, and USS *Vincennes*). In all 4 cases, better training or procedures could have prevented this failure and the consequent accident
10 All these accidents could have been avoided, most of them easily
11 Accidents may be normal but there are always opportunities to prevent any specific accident

APPENDIX TO CHAPTERS 5 AND 6

The Appendix contains four tables and their brief explanation, summarising the twenty-one accidents reviewed in Chapters 5 and 6.

DISCUSSION: TABLE 6.1 NATURE OF SYSTEM FAILURE

Each accident exemplifies a particular accident cause. The example cause (entered in column 3 of Table 1) is not chosen scientifically; it is chosen as one obvious cause, and one which is unlikely to be seriously disputed. Run your eye down column 3 to see whether you agree. It provides us with a list of known accident causes, a list of causes which we can address and seek to eliminate. So far, so good.

We have suggested three categories of system failure, three gates in our castle wall.

Self-Inflicted Injuries (column 4) are as common as we feared. Leaving aside the Comet accident where we are not in a position to make a judgement, this column shows us that there was an element of self-inflicted injury in every single one of the remaining twenty accidents, 95% of all cases.

The most common self-inflicted injury is complacency; it usually takes the form of inadequate consideration of a known risk. It is an obvious factor in eleven of the twenty-one accidents, 52% of our cases. In the R.101, the self-inflicted injury was primarily culpable ignorance; at Tenerife it was unnecessary haste; with the USS *Vincennes* it was culpably deficient planning.

Many accidents reveal more than one self-inflicted injury. Any self-inflicted injury indicates an avoidable accident. It has serious implications in the attribution of responsibility and blame.

Double Agents (column 5) are cases involving either over-reliance on safety measures, or of those measures introducing new risks.

Table 6.1 Nature of System Failure

Event	*Date*	*Failing Illustrated*	*System Defence Failure*		
			Self-Inflicted	*Double Agent*	*Shifting Sands*
1	2	3	4	5	6
US Mail	1920s	Risk Transfer	Yes		
Ford Pinto	30 May 1905	Profit	Yes		
Titanic	14 April 1912	Complacency	Yes	Yes	
R.101	04 October 1930	Ignorance	Yes		
DH Comet	1953–54	Innovation	Innovation		
DH Trident	18 June 1972	Design	Yes		
DC-10 Door	03 March 1974	Denial	Yes		
Challenger	28 January 1986	Pressonitis	Yes		Yes
Korean 747	22 December 1999	Culture	Yes		
Tenerife Collision	27 March 1977	Authority	Yes		
DC-10 Slats	25 May 1979	Improbability	Yes		Yes
Ocean Ranger	14 February 1982	Complacency	Yes	Yes	
Bhopal	03 December 1984	Self-Regulation	Yes		Yes
USS *Vincennes*	03 July 1988	Planning	Yes		
Piper Alpha	06 June 1988	Procedures	Yes		Possible
Lockerbie	22 December 1988	Priorities	Yes		Yes
Herald of Free Enterprise	06 March 1987	Direction	Yes		Yes
Valujet	11 May 1996	Overstretch	Yes		
Air France Concord	25 July 2000	Denial	Yes		Probable
NHS-QMC	03 January 2001	Culture	Yes		Yes
Twin Towers	11 September 2001	Inertia	Yes		

Column 5 is interesting, and probably misleading. It indicates the definite occurrence of defence failure by double agent in only two of twenty-one cases. It is likely that, in general, this form of failure is more frequent. For example it has contributed to more than one Airbus accident, when the pilots have mistakenly relied upon the protection of the clever but confusing autopilot system.

Shifting Sands (column 6) are cases where a mismatch can be identified between earlier intentions and later activity. It can be seen in the DC–10 slat accident (modified engine removal procedure), in the *Challenger* accident (increased acceptance of risk), at Bhopal (uncorrected drift into lax practices), in the *Herald of Free Enterprise*, Lockerbie and NHS-QMC events (in all of which there was knowing acceptance of lax practice). It was possibly a factor in the Piper-Alpha accident, and probably in the Air France Concord crash.

DISCUSSION: TABLE 6.2 LOCUS OF SYSTEM FAILURE

To identify the locus of the system failures, we sub-divide the system of operations into four functions: design, direction, management, and final operator.

These functions closely relate to chronological phases. Design precedes direction, which precedes management, which precedes the final operator. The earlier the phase of a failure, the more the phases which follow; arguably, therefore, the more the opportunities to rectify the initial failure.

Table 6.2 does not attempt to identify all system functions in which there may have been a failure. It records the functions in which the failures were most obvious, and which are unlikely to be seriously disputed. With this *caveat* we find that in twenty of the twenty-one accidents, failures occurred in more than one function or phase; in one, the Comet, we cannot make a judgement. This finding is evidence in support of the concept of an accident as a crack spreading across the system, a failure of the system defences.

The *Design Function* (column 4) was a locus of important failure in twelve accidents: the Pinto, *Titanic*, R.101, Comet, Trident, DC–10 (twice), *Challenger*, Ocean Ranger, *Herald*, Concord, 57% of cases. Without doubt, in each of these cases there were abundant opportunities subsequent to the Design Function to have rectified the failure.

By contrast, subsequent to a serious error by a *Final Operator*, as in the accident at Tenerife, there may be no opportunity for rectification. However, and importantly, there is nearly always opportunity *before*

Table 6.2 Locus of System Failure

Event	*Date*	*Failing Illustrated*	*Locus of Failed Function*			
			Design	*Direction*	*Manage-ment*	*Final Operator*
1	2	3	7	8	9	10
US Mail	1920s	Risk Transfer		Yes	Yes	
Ford Pinto	30 May 1905	Profit	Yes	Yes		
Titanic	14 April 1912	Complacency	Yes	Yes		Yes
R.101	04 October 1930	Ignorance	Yes	Yes	Yes	
DH Comet	1953–54	Innovation	Yes			
DH Trident	18 June 1972	Design	Yes		Yes	Yes
DC-10 Door	03 March 1974	Denial	Yes	Yes	Yes	
Challenger	28 January 1986	Pressonitis	Yes	Yes	Yes	
Korean 747	22 December 1999	Culture			Yes	Yes
Tenerife Collision	27 March 1977	Authority			Yes	Yes
DC-10 Slats	25 May 1979	Improbability	Yes	Yes	Yes	
Ocean Ranger	14 February 1982	Complacency	Yes	Yes	Yes	Yes
Bhopal	03 December 1984	Self-Regulation		Yes	Yes	
USS *Vincennes*	03 July 1988	Planning		Yes	Yes	Yes
Piper Alpha	06 June 1988	Procedures		Yes	Yes	
Lockerbie	22 December 1988	Priorities		Yes	Yes	
Herald of Free Enterprise	06 March 1987	Direction	Yes	Yes	Yes	Yes
Valujet	11 May 1996	Overstretch		Yes	Yes	
Air France Concord	25 July 2000	Denial	Yes	Yes	Yes	
NHS-QMC	03 January 2001	Culture	Yes	Yes	Yes	Yes
Twin Towers	11 September 2001	Inertia		Yes	Yews	

the event for good procedures to prevent the fatal error. Had KLM crew training emphasised that aircraft safety must be given priority over deference to authority, or had air traffic control been of a higher standard, there would have been no accident at Tenerife. The fact that there may be no recovery from the error of a final operator highlights the fact the need for well-constructed, firmly enforced operating procedures.

It is the responsibility of *the Direction and Management Functions* to construct and enforce these safe procedures. Leaving aside the Comet accident in which we are not in a position to judge, there was a failure of direction and/or management (columns 8 and 9), in every single one of the remaining twenty accidents. A failure of direction or management in 100% of the cases? A staggering indictment of the hierarchies.

And *Final Operator Failure* (Column 10)? In eight of twenty-one cases, in only four of which (*Titanic*, Tenerife, Korean 747, and *Herald of Free Enterprise*) could the final operator error be considered truly a culpable failure of the final operator. Four out of twenty-one. 19%.

Compare that 19% to the figure of 70–80% of accidents which are customarily attributed to the error of the final operator. And compare that 19% with the 100% attributable to failures of direction or management. Our analysis suggests that the final operator is the least of the culprits, a very long way behind the directors and managers. But the directors and managers nearly always survive the accident and, by one means or another, often exert a massive influence on the inquiry.

DISCUSSION: TABLE 6.3 NATURE OF DECISION FAILURE

We likened a risk management decision to a three-legged stool. It rests upon legs of information, intelligence, and integrity.

Information (column 11) is the data needed to take an informed decision. The deciding question is: was the necessary information available to the decision-makers? There are two failures in this column, the designers of the Comet and the final operators of the *Ocean Ranger* did not have the necessary information.

Table 6.3 Nature of Decision Failure

Event	*Date*	*Failing llustrated*	*System Defence Failure*		
			Info	*Intel-ligence*	*Integrity*
1	2	3	11	12	13
US Mail	1920s	Risk Transfer			Fail
Ford Pinto	30 May 1905	Profit			Fail
Titanic	14 April 1912	Complacency		Fail	
R.101	04 October 1930	Ignorance		Fail	
DH Comet	1953–54	Innovation	Fail		
DH Trident	18 June 1972	Design			Fail
DC-10 Door	03 March 1974	Denial			Fail
Challenger	28 January 1986	Pressonitis			Fail
Korean 747	22 December 1999	Culture		Fail	
Tenerife Collision	27 March 1977	Authority		Fail	
DC-10 Slats	25 May 1979	Improbability			Fail
Ocean Ranger	14 February 1982	Complacency	Fail		
Bhopal	03 December 1984	Self-Regulation			Fail
USS Vincennes	03 July 1988	Planning		Fail	
Piper Alpha	06 June 1988	Procedures		Fail	
Lockerbie	22 December 1988	Priorities			Fail
Herald of Free Enterprise	06 March 1987	Direction			Fail
Valujet	11 May 1996	Overstretch		Fail	
Air France Concord	25 July 2000	Denial			Fail
NHS-QMC	03 January 2001	Culture			Fail
Twin Towers	11 September 2001	Inertia		Fail	

Intelligence (column 12) is the ability of the decision-makers to process the information adequately. It covers their knowledge, expertise and intelligence. The deciding question is: were they capable of processing the information adequately? There are eight failures in this column.

Integrity (column 6) is the maintenance by the decision-makers of the priority of the imperative of accident avoidance, and thus their rejection of other interests: haste, expedience, profit, personal benefit, etc. The question is: did they subordinate the imperative of accident avoidance to other interests? Answer: Yes, too often, in eleven of the twenty-one cases. This is the most common reason for decision failure.

DISCUSSION: TABLE 6.4 OVERVIEW OF SYSTEM FAILURE

Table 6.4 is the amalgam of Tables 1, 2 and 3.

Table 6.4 Overview of System Failure

Event	*Date*	*Failing Illus-trated*	*Nature of Failure*			*Locus of Failure*				*Decision Failure*		
			Self-Inflicted	*Double Agent*	*Shifting Sand*	*Design*	*Direction*	*Manage-ment*	*Final Op*	*Info*	*Intel-ligence*	*Integrity*
1	2	3	4	5	6	7	8	9	10	11	12	13
US Mail	1920s	Risk Transfer	Yes				Yes	Yes				Fail
Ford Pinto	30 May 1905	Profit	Yes			Yes	Yes					Fail
Titanic	14 April 1912	Complacency	Yes	Yes		Yes	Yes		Yes		Fail	
R.101	04 October 1930	Ignorance	Yes			Yes	Yes	Yes			Fail	
DH Comet	1953–54	Innovation	Innovation			Yes				Fail		
DH Trident	18 June 1972	Design	Yes			Yes		Yes	Yes			Fail
DC-10 Door	03 March 1974	Denial	Yes			Yes	Yes	Yes				Fail
Challenger	28 January 1986	Pressonitis	Yes		Yes	Yes	Yes	Yes				Fail
Korean 747	22 December 1999	Culture	Yes					Yes	Yes		Fail	
Tenerife	27 March 1977	Authority	Yes					Yes	Yes		Fail	
DC-10 Slats	25 May 1979	Improbability	Yes		Yes	Yes	Yes	Yes				Fail
Ocean Ranger	14 February 1982	Complacency	Yes	Yes		Yes	Yes	Yes	Yes	Fail		
Bhopal	03 December 1984	Self-Regulation	Yes		Possible	Yes	Yes	Yes				Fail

Table 6.4 Overview of System Failure (continued)

Event	*Date*	*Failing Illus-trated*	*Nature of Failure*			*Locus of Failure*				*Decision Failure*		
			Self-Inflicted	*Double Agent*	*Shifting Sand*	*Design*	*Direction*	*Manage-ment*	*Final Op*	*Info*	*Intel-ligence*	*Integrity*
1	2	3	4	5	6	7	8	9	10	11	12	13
USS *Vincennes*	03 July 1988	Planning	Yes				Yes	Yes	Yes		Fail	
Piper Alpha	06 June 1988	Procedures	Yes		Possible		Yes	Yes			Fail	
Lockerbie	22 December 1988	Priorities	Yes		Yes		Yes	Yes				Fail
Herald/Fr/ Enterprise	06 March 1987	Direction	Yes		Yes	Yes	Yes	Yes	Yes			Fail
Valujet	11 May 1996	Economy	Yes				Yes	Yes			Fail	
Air France Concord	25 July 2000	Denial	Yes		Probable	Yes	Yes	Yes				Fail
NHS-QMC	03 January 2001	Culture	Yes		Yes		Yes	Yes	Yes			Fail
Twin Towers	11 September 2001	Inertia	Yes				Yes				Fail	

CHAPTER 7

ACCIDENT INVESTIGATION SHOULD RESULT IN ACCIDENT PREVENTION

IN THIS CHAPTER

It is doubtful that any organisation is specifically charged with, and properly funded for, the task of accident prevention. This includes the accident investigation agencies; they have the best opportunity for accident prevention, for two reasons:

First, accidents have to be investigated. Therefore there are always substantial funds available for a professional, searching investigation. By contrast, funds for accident prevention are extremely difficult to extract.[48] Second, investigation lays the foundation for prevention. A failed or flawed investigation is therefore an accident prevention opportunity missed, one of the many good reasons for which the integrity of the investigation process must be assured. Unfortunately there are powerful forces opposing integrity and objectivity; we will identify them.

The two major investigations reviewed in this chapter both failed as tools of accident prevention, and for diametrically opposite reasons.

- Scott Snook's *Friendly Fire* investigates a US military tragedy in Iraq. It is an exemplary study. Snook concludes that the cause of the event remains uncertain. We respectfully disagree.

48 What, you wish to study the possibility of a slat failure coinciding with an engine failure coinciding with a take-off? Bit far-fetched, old boy. Do me a favour, run along now.

- The RAF inquiry into the crash of Chinook ZD 576 on the Mull of Kintyre in 1994 is also exemplary, but for the wrong reasons. It is a textbook example of a monumentally flawed inquiry

We examine each of them carefully because they provide us with important lessons.

- The causes of both accidents were deeply rooted in their systems of operation. Although the accidents occurred in seconds, their roots had lain in place, uncorrected, for years
- These accidents constitute powerful supporting evidence for our key concept that accidents result from the failure of the system of operation

First we discuss some important considerations which underlie accident investigation. The need for closure drives the investigation process. Questions of responsibility and blame seriously impede the use of investigation as a tool of prevention. The investigators' role is to establish what happened and why it happened. They are not normally directed actively to investigate the prevention of further similar accidents.

THERE MUST BE CLOSURE

In Chapter 1 we touched on the human need to make sense of the world. We become deeply unhappy if we feel that we face the inexplicable. Hence our hard-wired need to explain every event by finding, or theorising, or fabricating, a cause for it. This need underpins our search for closure after a traumatic experience. Closure is achieved by putting the mind at rest, so that it is no longer dominated by the trauma but is freed up to operate effectively again. Closure is important. It enables life to go on.

If the experience is not serious but merely unpleasant, we can achieve closure by making a joke of it. Laughter is the easiest and greatest stress-buster.[49] For that reason it is prevalent in the most stressful or harrowing professions. Undertakers tend to be highly

49 And, even better, it does not involve the NHS.

amusing. And it is most unusual to read a soldier's reminiscences without falling off the sofa at the best jokes. (I am not talking about senior officers' memoirs; their hilarity is generally unintentional.) There is also a good deal of merriment amongst professionals in aircraft operation. The downside of this (built-in-but) is that newcomers and the less committed sometimes fail to observe that the merriment is strictly subordinated to the demands of the situation. Avoid aircraft operated by newcomers or the less committed.

When one of life's truly horrible experiences is involved, a major accident or bereavement, putting it to rest with laughter is out of the question. Closure is difficult, perhaps impossible, unless some reason for the event can be found and accepted. To our cause-and-effect minds, the reason for the accident is the accident cause. Interestingly, our innate preference for an easily understood explanation translates into a search for a simple, preferably single, cause, the failure of a person or of a component.[50] The failure of the system of operation would be a verdict too nebulous to enable closure.

Conversely, a simple, single cause is inadequate as an investigation verdict. The system of operation has failed; it has not achieved its Type B objective, the avoidance of an extreme adverse event (Chapter 2). Therefore the investigation must not stop at a single cause. It must go on to identify all the points of failure of the system of operation. This further assists closure. Perhaps more important, it enables system deficiencies to be rectified, and thus similar accidents to be prevented. Any investigation which does not achieve this has failed.

RESPONSIBILITY AND BLAME

It is important to distinguish responsibility from blame. By our definition:

- Responsibility is: did you do it?
- Blame is: were you at fault in doing it?

50 Hence, perhaps, the superficial attraction of root cause analysis.

Responsibility can be established with a degree of objectivity. Were you or were you not driving that car? Did you or did you not carry out that order? Were you or were you not supervising that operation?

Blame is less straightforward. It arises from a deficiency in the discharge of duty, whether duty in a military context or duty of care in a civil context. Deficiency is more judgemental than responsibility. It may be attributed to an act of commission: something you did, or to an act of omission: something you omitted to do. Central to any assessment of deficiency is the question of capability: were you capable of taking the desired action? Therefore blame has four preconditions:

1 Responsibility
2 Duty
3 Capability
4 Deficiency

All four have to be established before any case for blame can be made.

It is unfortunate that, in general usage, responsibility and blame are often used synonymously. 'You were responsible for the accident' equals 'you were to blame for the accident'. In our more careful usage they represent significantly different concepts.

BLAME

Blame carries a burden which may be extremely serious to persons or to organisations. It signifies liability with its implications of huge costs in legal damages and legal costs (the euphemism for lawyers' hefty fees). It implies incompetence. It can destroy reputations. It may result in dismissal or in career destruction. It can destroy self-respect. All of which adds up to a massive potential cost. For that reason individuals and organisations may well fiercely resist the attribution of blame to themselves.

The identification of the accident cause may not be enough for closure to be achieved. A further step may be needed: the identification of the reason underlying the cause. The reason easiest to understand and therefore easiest to accept is that someone is to blame. Hence the great pressure to identify a person at fault. Further pressure comes from those who are exonerated by the identification of fault in another. Finding someone guilty finds everyone else not guilty. It closes the inquiry. Very convenient.

'Skunk Airways announces that its investigation of the accident has determined that the prime cause was Captain Scapegoat's failure to achieve the airline's normal high standards; Captain Scapegoat has therefore been dismissed.' Sub-text, there was nothing wrong with the management of the airline, responsibility and blame for the accident was attributable solely to one individual, Scapegoat, formerly Captain.[51]

This mechanism, scapegoating, is not confined to accident investigation. You can see it at work whenever something serious goes wrong, and whenever something becomes serious by becoming public knowledge. At government level, ministers are encouraged to write resignation letters admitting that they were not up to the job (an excellent cover story because it is usually true); this takes the heat off the Prime Minister for his latest disaster. Failing that, the news of their utter inadequacy is leaked to the media by their delightful erstwhile friends. Both amount to the same thing, the egg is off senior faces. In the 1980s, British Airways carried out a massive cull of managers, inevitably terminating some excellent people as well as many less excellent. The mantra was that if it happened on your watch, then you were to blame. If you knew it was happening, you should have stopped it. If you did not know it was happening, you should have known. Either way, goodbye. But when one of the airline's competitors made a public issue of some supposedly unfair BA policy, the mantra was

51 One of my RAF chums, Captain Sam Shaw, was flying for a then emergent, now respectable, British airline. He had flown his normal maximum duty day, say 12 hours. The chief pilot then ordered him to fly two more sectors, ending in Moscow, well beyond all flight time limits. Sam refused. The chief pilot said fly or report to my office. Sam was an Ulsterman, a breed whose babies chew on the red flag of revolution for comfort. So he arrived at the chief pilot's office and gave him a blunt reminder of the law of the land. Are pilots still put under pressure to break the law? You bet.

quickly forgotten. Neither Lord King nor the then Sir Colin Marshall were to blame. Because they did not know it was happening.[52]

The primary task of an objective, honest, investigation is to establish the facts. For this the cooperation of witnesses is very important. For that reason the best investigation agencies strenuously seek, and some are specifically charged, to avoid any attribution of fault. If the possibility of fault and blame has to be investigated, it must be the subject of a separate inquiry.

This book seeks to inform its readers of the realities of accident causation. They will then be able to make their own informed judgements on the matters of responsibility and blame for accidents. They will also be in a position to put the boot into misleading statements. And, if they so wish, into those who seek to mislead.

FROM INVESTIGATION TO PREVENTION

The tasks of investigation are to establish:

1. What happened
2. Why it happened
3. How it might have been prevented, and thus how similar accidents might be prevented in the future

In practice the best investigation agencies do the first two tasks very well, but they make limited inroads into the third, accident prevention. There are two reasons for this.

First, accident prevention requires the identification of points of responsibility. This is coincidentally the first step in the attribution of blame. As we have already seen, the priority of investigation agencies is to obtain all possible facts pertaining to the accident. A requirement to identify the points of responsibility may inhibit the cooperation of witnesses. For this reason, agencies tend to draw back from such identification.

52 The employment contract: You do not pass the buck up to the top brass. And they do not pass the big bucks down to you. Seems fair.

Second, the regulation of operations is the task of the regulatory agency, not the investigative agency. In Britain air accident investigation is in the hands of the AAIB (Air Accident Investigation Branch), and regulation is in the hands of the CAA (Civil Aviation Authority). The corresponding organisations in the USA are the NTSB (National Transportation Safety Board) and the FAA (Federal Aviation Agency). Friction between the investigation and regulation agencies is the norm. The investigators feel that the regulators are too feeble in their imposition of safety measures upon the industry. However the regulators are saddled with two mutually conflicting objectives, the nurturing of the industry versus the imposition of (expensive) safety. As we have seen in the previous two chapters, the FAA has been regularly, seriously, and rightly criticised for its failures in its implementation of its second objective. In GB, the CAA has also been accused of dragging its feet, justifiably in the view of many observers. But whereas the FAA subordinates safety to that Great American God, the imperative of commerce, the CAA subordinates safety to the civil servants' preference for cooperation over coercion. Whether this is for reasons of its restricted budget, or of the mandarins' exquisite courtesy, or of the attractions of a quiet life, you may decide for yourself.[53]

The international agency, the International Civil Aviation Organisation, specifies that accident prevention is the primary aim of accident investigation. The fact remains that, in most countries, neither the regulatory nor the investigation agency dedicates itself single-mindedly to accident prevention. Mind how you fly.

FRIENDLY FIRE

Friendly fire: irony or euphemism? Whichever you prefer, it seems that friendly fire is a built-in-but which closely accompanies the US forces. Apologists refer to it as an accident of war, as though that closes the matter. It does not. Accidents can be prevented, and that includes accidents of war.

53 Mandarin: old fashioned term for a very senior civil servant. Not to be confused with mandarin, a wrinkled fruit.

I have asked RAF pilots who flew in the Second World War for their views on friendly fire. In summary, they all considered it to be excusable only very, very rarely. Peter Brothers, distinguished fighter leader and Battle of Britain Hurricane ace, told me about two incidents both of which he regarded as incompetence. The first occurred when he was operating out of France early in the war. He saw with pleasure that his No.2 was joining up on him after a dogfight. He was less pleased when the No.2 started blazing away at him from close range, just as an Me 109 had done seconds before. Peter took violent evasive action. Words were exchanged. He then sent the No.2 back to gunnery school, a considerable humiliation, for missing from close range. In the second incident he was attacked by a Spitfire who also missed from an easy range. This Spitfire then claimed him as an Me 109 destroyed. The attacking squadron of Spitfires was known for being enthusiastic but disorganised, so nothing more was said. Ian Ponsford, wartime Spitfire pilot, said that he knew of only one case. In this, an RAF Typhoon was landing at a temporary airfield in Belgium. It was shot down by a USAAF P–47 Thunderbolt and the RAF pilot was killed. The US pilot was court-martialled. Ian was an eminent lawyer; he was measured in his words both by nature and by profession. He regarded the accident as entirely avoidable and therefore unforgivable.

Culpability can be assessed only when an accident is investigated in detail and in context. An important aspect of context is the degree of danger faced at the time by the perpetrator. If he is within a split second of losing his own life, then he has no time in which to weigh his decision to fire. In too many recent cases US ground-attack pilots have been in very little danger when they opened fire and killed British soldiers. In the case which we will now review, the two US fighter pilots were effectively in no danger.

To accept that friendly fire accidents of war are inevitable is the worst possible approach to their prevention.

TRAGEDY IN THE SKIES: TWO BLACK HAWKS DOWN

Scott Snook's book, *Friendly Fire*, is an accomplished analysis of the incident in which two US Air Force F–15 fighters shot down two US Army Black Hawk helicopters, killing all 26 crew and UN peacekeepers on board (14 April 1994). In Chapter 2 we used the accident as an example of shifting sands; common practice had drifted, or shifted, a long way from the original plan. Snook gave this the descriptive name 'practical drift'. It is one of the many forms of shifting sands.

In April 1991 Iraq accepted the UN ceasefire conditions, officially ending the first Gulf War. A massive humanitarian effort was launched to help the huge number of Kurdish refugees in northern Iraq. The most immediate concern of Operation Provide Comfort was to 'stop the dying and stabilise the situation'.[54] A no-fly zone, forbidden to Iraqi aircraft, was established in N Iraq. The daily routine was for an AWACS (Airborne Warning And Control System) aircraft to fly into position to control all coalition aircraft in the zone, then for the US Air Force to carry out a fighter sweep to ensure the zone was clear of hostile aircraft, then for coalition aircraft and forces to enter the zone. There was little threat to their operations from the Iraqis who 'had never violated the nofly zone'.

The Event

14 April 1994

0845hr: the AWACS was in position.

0935hr: two US Army Black Hawk helicopters reported their entry into the zone to the AWACS.

0954hr: the Black Hawks informed the AWACS of their next routing.

1020hr: two US Air Force F–15 fighters reported to the AWACS that they were entering N Iraq.

54 There is a dreadful irony in the name Provide Comfort and its good intention and the present (2008) reality. Good intentions pave the road to hell, they do not pave the road to democracy. Hell is where the Iraqis are now living, that is those who are still living.

1022hr: the F–15s reported 'radar contact on a low-flying, slow-moving aircraft'. Shortly afterwards they identified these aircraft as Hinds (Russian helicopters used by the Iraqi Air Force). The AWACS acknowledged.

1022–30hr: there were six communications between the F–15s and the AWACS.

1030hr: the F–15s destroyed the two Black Hawks with two missiles, killing all twenty-six on board.

The event can be summarised from the points of view of the Black Hawks, F–15s, and AWACS.

Black Hawks

- US Army helicopters had for at least two years routinely been entering the no-fly zone before the fighter sweep
- The Black Hawks kept the AWACS properly informed of their activities
- The Black Hawks were operating their IFF/transponders on the wrong code. (Transponders enable friendly aircraft to be distinguished from hostile aircraft, hence IFF: Identification Friend or Foe.)
- Black Hawk operations had used this wrong code for two years. They were not aware of their error. The AWACS were aware of the error but had not corrected it. The F–15s were not aware of it

F–15s

- The F–15s kept the AWACS properly informed of their activities
- The F–15s were not informed of the presence of the Black Hawks in the no-fly zone
- The F–15 leader referred to his aircraft recognition booklet to confirm his visual identification of the helicopters as Hinds
- The F–15s attempted to confirm the helicopters' identity by interrogating their transponders; this was partially impeded by terrain as the helicopters were flying low in valleys

AWACS

- Although nominally in control, by custom and practice the AWACS deferred to the fighters
- The AWACS did not take any of its several opportunities to inform the F–15s of the presence of the Black Hawks
- No individual took control of the disorganised AWACS operation and imposed order and efficiency

Major System Failures

- The AWACS was not tasked to monitor friendly helicopter traffic in the no-fly zone (according to an AWACS officer after the accident)
- The entry of the helicopters into the no-fly zone, which had been routine for two years, was not formally required to be recorded in the system. Therefore this fact was not conveyed to the F–15s in their briefing
- The F–15s and the Black Hawks were equipped with incompatible radio systems. They could not communicate with each other
- The use of the wrong IFF code by the Black Hawks had gone uncorrected for at least two years

System Failures at the Direction Level

- All the above major system failures are attributable to failures of command and control. All could have and should have been corrected by those responsible for command and control
- The AWACS operation was a total mess. The crew was *ad hoc*, not experienced at working with each other. The AWACS in-flight procedures were a mess. No one was responsible for tracking friendly helicopters. The fighters were not under control
- AWACS crews were tired and jaded. A morale survey conducted amongst them shortly before this event had yielded savage responses, including a series of blunt warnings that an accident was imminent. 'It will take a serious tragedy, an airplane and crew lost

…, before they in their ivory tower wake up to the realities of AWACS today'

- The integration of the operations of the US Air Force with those of the US Army was abysmal
- The role of F–15s was primarily medium- and high-level interception. The pilots were not trained in low-level operation

Failures of the F–15s

As ever, pilots are the crackstoppers of last resort. If an accident occurs the question of whether the pilot could have prevented it always arises. Either F–15 pilot could have prevented this tragedy.

There were strong extenuating circumstances for the fighters: the failure of the AWACS to inform them of the Black Hawks, the Black Hawks' selection of the wrong IFF code, and the poor information in the pilots' aircraft recognition guide (the Black Hawks had many of the supposed distinguishing features of the Hind). Further, the F–15 pilots, in their normal role as medium/high-level fighter pilots, may well fire missiles at targets which they have identified only electronically, and not visually. To put them without appropriate training into the different role of low-level operations is reprehensible. They were evidently somewhat nervous at finding themselves in valleys surrounded by mountains. Nonetheless before anyone fires a gun or launches a lethal weapon, it is no more than common-sense to make a final positive check before firing that the right target has been selected. A few more seconds' deliberation or a closer visual inspection would have revealed that the Black Hawks' paint scheme was very different from that of Iraqi Hinds.

The Cause of the Tragedy

Snook in his conclusions gives the impression that he was looking for a smoking gun, a clear and simple cause, even a root cause.

> There weren't any bad guys; hence no one to blame … There weren't any catastrophic failures of material or equipment; hence

> nothing to fix ... I found nothing that I couldn't explain, given what we already know about individual and group behaviour in complex organisations ... I could find nothing extraordinary ... If no one did anything wrong ... no unexplainable surprises ... if nothing was abnormal ... ., then what have we learned?

Here we part company with the admirable Snook. Yes, every part of the tragedy can be explained. But that is not the point. Nor does it excuse it. Underlying this preventable tragedy were two major failings, the failings of the command and control system which set the context for a probable extreme adverse event, and the failure by several individuals to intervene in the parlous scene and to prevent the accident. Snook should not be looking for a smoking gun, there isn't one.

And we most certainly *have* learned. We learned that there was a massive failure of the system of operation, its orders, its procedures, and its deployment of personnel. The crack ran right through the system; the major failures had sat there for two years, uncorrected. In looking for bad guys and for catastrophic failures of materials and equipment, Snook is adopting the classic approach. In looking for a simple accident cause, he is looking for the wrong thing.

The cause of the tragedy is very clear. The system was not fit for purpose. The failure of the command and control bureaucracy, together with the failures of individuals to intervene, constitute a total system failure. And if there were no bad guys, there were certainly some who were not very good. We can fully understand Snook's reluctance to criticise members of our armed forces, and we share it. But we are here to prevent accidents.

Snook says that the accident was normal in Perrow's sense (normal in a poorly insulated, incomprehensible system). We can agree with that.

He says it was also normal in the sense that 'it occurred as the result of normal people behaving in normal ways in normal organisations'. We cannot agree with that. The US Air Force is not a normal organisation. It is an extreme killing organisation. It must be particularly mindful of its Objective B, preventing extreme adverse events in the wrong place. Accidents will be the *normal result* in such an organisation when procedures are poor, training is poor and motivation is poor.

In the best of organisations accidents *may* happen. In other organisations, they *will* happen. Let no-one say that it is normal to have easily avoidable accidents.

Scott Snook's book is not a routine accident investigation. It is a masterly research study. Snook seeks to understand the event at several levels and to synthesize these microstudies into a macrostudy of the event as a whole. His own record as a combat officer facilitated his access to classified material and it enabled him to gain the confidence of the doubtless horrified personnel involved. The US government and the US Air Force made strenuous efforts to understand how such an event could occur. The 27-volume accident report, the 2600-page record of a court martial, and the Special Congressional Hearing were the tip of the massive investigation iceberg. To this material Snook had access.

The commitment to openness of the US authorities is enormously to their credit. It reveals the US determination to identify deficiencies in the US Air Force, even when they may be culpable.

We will return to this event in our next chapter, where we discuss the prevention of accidents.

SCANDAL IN THE SKIES: ACCIDENT TO RAF CHINOOK ZD 576

The Event

2 June 1994

- A new mark of RAF Chinook, the HC2, was tasked to fly twenty-five of the country's most important counter-terrorist officers from N Ireland to Fort George, near Inverness.
- The RAF's own expert test pilots had had a number of serious incidents with the HC2. They had refused to certify it fit for operational service. They had refused to fly it themselves until problems with the software had been resolved.

- The weather was marginal; low cloud was a possibility
- The route was a short sea-crossing followed by flight between mountains
- The first waypoint was the cliff-face on the Mull of Kintyre
- The HC2 was not permitted to climb above the mountains by its icing limit (the possibility of aircraft icing)
- The Captain asked for an earlier mark of Chinook, the HC1, which could have climbed over the mountains. His request was refused
- The squadron's earlier request to breach HC2 limits in case of necessity had been refused

RAF Chinook ZD 576 crashed into the mountainside on the Mull of Kintyre. All on board were killed.

The Board of Inquiry

RAF air accidents are investigated by an RAF Board of Inquiry (BoI), established and controlled by the RAF Command involved. The members of the BoI are RAF serving officers.

For expert advice, they can call on a professional investigator from the civil AAIB. Their official report is transmitted up the command chain to two air officers.

The Investigation

The Chinook struck a steep incline. It bounced three times, disintegrating into small pieces, which were then consumed in a fireball. The expert accident investigator from the AAIB described the evidence as 'thin'. He would have been looking first for evidence of a structural failure or an engine failure. Such evidence might be expected to survive. Next he would be looking for evidence of an engine or flight control malfunction; such evidence would be in the form of marks or damage in the engine, or in the appropriate control system, or contamination of hydraulic systems, or traces of electronic malfunction. Given the massive damage to the helicopter, the detection of evidence of minute marks, or of microscopic fragments in hydraulic fluid, would be difficult in the extreme. A transient malfunction or an electronic malfunction may leave no trace at all, apart from a wrecked aeroplane.

The Findings of Cause

The Board concluded:

> There were many potential causes of the accident and despite detailed and in depth analysis the Board was unable to determine a definite cause ... The Board concluded that the most probable cause was that the crew selected an inappropriate Rate of Climb to safely overfly the Mull of Kintyre.

The first senior reviewing officer stated: 'It is incomprehensible why two, trusted, experienced and skilled pilots should ... have flown a serviceable aircraft into cloud-covered high ground.' He then went on to surmise:

> In my judgement none of the possible factors and scenarios (postulated by the Board) are so strong that they would have been likely to prevent such an experienced crew from maintaining safe flight ... I am ... forced to conclude that [the captain] was negligent to a gross degree.

The second senior reviewing officer stated:

> Without the irrefutable evidence which is provided by [a black box and cockpit voice recorder] – neither of which was fitted to the HC2, there is inevitably a degree of speculation.

He then went on to state with apparent certainty: 'the actions of the pilots were the direct cause of the crash ... this amounted to gross negligence.'

RAF Findings of Pilot Error

The RAF has a scale of human culpability. Error of judgement is an honest mistake with no lack of zeal. Negligence justifies severe criticism; it may be blameworthy to a minor degree or to a gross degree. The gravest condemnation is recklessness or disobedience.[55] Gross negligence is a very severe verdict, probably ending a pilot's flying and RAF careers.

The readiness with which the RAF has blamed its pilots for flying accidents has been criticised for decades. It is particularly resented when the pilot is killed in the accident; he is then unable to recount what happened or to defend his actions. He is posthumously condemned by summary diktat, and on the personal judgement of senior officers with personal vested interests, and who were not there. Worse, a finding of negligence is likely seriously to reduce the compensation paid to his widow, perhaps halving it. As a result of past criticisms RAF regulations stated unequivocally, 'Only in cases in which there is absolutely no doubt whatsoever should deceased aircrew be found negligent.'

Earlier Endeavours to Prevent Easy Verdicts of Pilot Error

Thoughtful aviation circles have long been fighting the injustice of easy pilot error verdicts.

In 1959, the Minister of Aviation established a committee under the Hon. Mr Justice Cairns to enquire into the law and practice governing the investigation of civil aircraft accidents.

The Cairns Report:

- cited 'the tendency to blame an individual rather than an organisation or a piece of equipment'
- gave the right to request a judicial review chaired by a QC to any person or body having objections to proposed findings of an accident investigation
- formalised the rights of any person to whom the Inspector intended to attribute blame

In 1982, the RAF established an RAF Working Party under Air Cdr Derek Hine to examine Boards of Inquiry into flying accidents. It

55 As for instance trying to fly under a bridge and hitting it. Bad flying. Or, doing it so slowly that someone is able to take the aircraft registration. Stupid flying.

operated under Condor Rules, rules of confidentiality which allowed officers to speak their minds of the RAF establishment and of their superior officers without committing career suicide.

The criticism of the RAF 'verdict' on ZD 576 was widespread and vociferous. This led, in 2002, to the establishment of a Select Committee of the House of Lords to review the verdict. Air Cdr Derek Hine reported the findings of his 1982 Working Party to the Committee.

> ... there was strong dissatisfaction amongst aircrew that they were being unfairly blamed ... The Working Party found consistent criticism from aircrew and groundcrew that BoIs were flawed by being convened and controlled, by the very same officers who had command and control of the flying operations under investigation. ... disturbing evidence was produced demonstrating that at least two Boards had been corrupted by Higher Authority and pressured while the Boards were still deliberating.

The RAF regulation 'Only in cases when there is absolutely no doubt whatsoever should deceased aircrew be found negligent' was drafted by the Hine Working Party. Air Cdr Hine emphasised to the Select Committee both the clarity of its intention and its mandatory status:

> The Regulation ... 'demands a high degree of proof. That was the intention ... This was universally agreed by the Air Force Board and all Air Force Departments ... It was the intention of the Air Force Board et al that this recommendation should be set in stone and ... followed by all involved in the BoI process – ie the Board (under its President) and ALL subsequent stages of the Higher Authority process.'

Derek Hine then noted 'a variety of misleading statements to the public attributed to MoD'. This included the startling statement that '*the conditions for negligence applied only to the Board under its President, not to Higher Authority*' (*emphasis* in original).

In 1986 continued disquiet with RAF Boards of Inquiry led the Minister for State for Defence Support to ask Bill Tench, Chief Inspector of the civil Air Accidents Investigation Branch, to investigate the methods of conducting inquiries into military aircraft accidents.

Tench singled out the RAF from the three services for blunt criticism:

> The pervasive nature of the involvement of some Station Commanders, AOCs, senior staff officers and even Commanders in Chief, is an unwelcome intrusion upon what should be the complete independence of the Board of Inquiry.

In 2000 Bill Tench wrote to the father of one of the pilots killed in ZD 576, referring to the above quotation, 'It appears that this was completely ignored in respect of the ZD 576 inquiry.'

Thus since 1959 (at least) there has been recognition at official level of the dangers of easy pilot error verdicts. Erroneous verdicts are a double disgrace to the service. They are an appalling injustice, and they do nothing to prevent the next accident.

RAF Standard of Proof for Negligence

The RAF's mandated standard of proof for negligence is unequivocal, 'only in cases where there is absolutely no doubt whatsoever'. This equates to absolute certainty. In practical terms, absolute certainty can be attained only when there is:

- more than one independent source of positive, verifiable evidence
- no conflicting evidence
- no alternative explanation

In the accident to ZD 576 there was:

- no positive, verifiable evidence that pilot negligence had been the cause of the accident
- some conflicting evidence

- at least three alternative explanations, all arguably more probable than pilot negligence

This clearly makes it impossible to meet the RAF's own required standard of proof for negligence. So how did the RAF achieve the impossible? Please don't ask me. Ask the RAF.

The RAF Verdict

The Board's verdict that the most probable cause was that the crew selected an inappropriate Rate of Climb is one small step away from the finding that the aircraft hit a cliff. It tells us that the Board was unable to identify the accident cause. This is to their credit. The evidence was too 'thin' to support any positive finding.

The air officers gathered no more evidence, they called no new witnesses. On the basis of the same 'thin' evidence they pronounced their personal opinion on the accident cause. It was not, they opined, negligence but gross negligence, gross dereliction of duty, effectively a verdict of manslaughter.

Nowhere in Queens Regulations are the air officers empowered to substitute their personal opinion for the Board's verdict without meeting the RAF's mandated standard of proof. Amazingly, that is exactly what they did. Their unsubstantiated opinion became the RAF's official 'verdict'.

A Benchmark of Bad Practice

Air accident investigation is a complex task for professionals. It is preposterous to expect amateurs to become instantly proficient in its undertaking simply because they are RAF officers.

Accidents are caused by a failure of the system of operation. To entrust the investigation of the failure of the system to those responsible for its safe operation is plain stupid. It is unfair to them, and it is unfair to all involved. It is the equivalent of asking a judge to try himself for incompetence. The gravest doubts on his judgement will be inevitable. If he finds himself not guilty, let alone without giving his reasons, then dark suspicions will be unavoidable.

Independent experts of all callings joined together to greet the official RAF 'verdict' with a loud collective raspberry. Eminent lawyers, pilots, aviation engineers, computer professionals and accident investigators all found that, in their particular area of expertise, the verdict was seriously flawed.

Air officers are not used to having their judgements scrutinised. Unluckily for them, the ZD 576 Board of Inquiry process coincided with the open government initiative of the then Prime Minister, John Major. Suddenly and unprecedentedly their conduct was in the public domain.

The RAF refused to accept the objective verdict of the distinguished Select Committee of the House of Lords (a Lord of Appeal, three eminent QCs and one eminent engineer). The Select Committee stated, 'we unanimously conclude that the reviewing officers [the air officers] were not justified in finding that negligence on the part of the pilots caused the aircraft to crash.' The MoD was unabashed. 'The Ministry of Defence notes the Committee's report but does not accept its conclusion.' The RAF's answer in so many words was, we are the only judge of the matter and we judge ourselves right. The New Labour government was complicit in the injustice; it refused to order the RAF to withdraw its verdict.

Lord Craig of Radley, former Marshal of the RAF and Chief of Defence Staff, spoke in the Lords of the infallibility of air officers:

> Commanders must be respected and trusted by their subordinates, whether in matters of professional conduct or discipline. It is inappropriate to set up an inquiry into the professional judgement of the Air Marshals. At the end of the day, it is their professional judgement on which we must rely …

So air officers are entitled to unquestioning deference to their judgements. They answer to no one.

And if two young RAF pilots are ordered to fly an aircraft unsuitable for the route and conditions, and are killed on that duty, what, we may ask, happens to that respect and trust when Air Marshals judge it right to crucify the reputations of the pilots, judge it right to exonerate themselves, and judge themselves so infallible that they are unwilling to submit their judgement to professional scrutiny? Someone should tell them what happens to it.

A Contrast in Investigations

The investigations which followed the US incident of friendly fire were thorough and open. The inquiry and aftermath of the accident to ZD 576 were neither thorough nor open. Potential recruits to British armed forces should be warned that if the friendly fire of the US Air Force doesn't get you, then the friendly fire of your own senior officers and New Labour will.

Sign here.

Safeguarding the Investigation Process

The prerequisites of investigation are the integrity and objectivity of those directing it and of those conducting it.

The most common fault line lies in the direction of the investigation. The concept of an accident as a failure of the system of operation is now widely accepted. Therefore the need to exclude from the investigation process those with potential responsibility for the accident is no more than common-sense. And it is no more than common-sense to place the investigation in the hands of those with the necessary expertise. There is no better example of these imperatives than the RAF inquiry into the accident to Chinook ZD 576. This inquiry was conducted by amateurs. It was directed and controlled by those responsible for the system which failed. It did not determine what happened, why it happened, or how to prevent a recurrence. The RAF process following the accident to Chinook ZD 576 sets a benchmark for the comprehensive failure of an accident inquiry.

More subtle misdirection is achieved by the mandarins of the civil service. The careful framing of the terms of reference can exclude areas of deficiency from scrutiny. The careful choice of ambitious or malleable or useless or establishment figures to chair or to participate in inquiries can be a considerable influence on their course and outcome.[56] Even better if he is a distinguished personage of undoubted integrity. Alternatively the appointment of a celebrity with limited intellectual fire-power as chairman will assure his dependence on civil service guidance whilst also assuring the ready acceptance of his pronouncements by the public.[57]

The integrity and objectivity of professional investigators is rarely to be questioned in open and competent societies. This is not to say that their reports are immune from editing by their superiors nor, in some countries, from government pressure.[58] Our own AAIB often assists other nations in their investigations. Such reports are very diplomatic in their contents. Some years ago I wrote an acerbic comment on an investigation which notably omitted any mention of the several deficiencies of an African country's airfield. I received an angry phone-call from the investigator. 'I did mention them.' 'No, you did not.' 'Yes, I did.' 'No, you did not.' 'Yes, I did, but the boss removed the paragraph.' In such cases the omission of public criticism may well facilitate the rectification of deficiencies, and is therefore justified. Nonetheless, hmm.

ACCIDENT INVESTIGATION SHOULD RESULT IN ACCIDENT PREVENTION

Earlier in this chapter I suggested that the tasks of any investigation into any adverse event are to establish:

1 what happened
2 why it happened
3 how it might have been prevented, and thus how similar accidents might be prevented in the future

56 For a successful conclusion to this matter, we will be obviously thinking of a peerage, old boy.

57 As I found in 1974 when I tried to convince Douglas Bader's Committee that airline pilots were widely and dangerously fatigued.

58 I will not name names. *Absolument non.*

It is obvious, is it not, that the determination of *what happened* is the key task of any investigation. Without this prerequisite, there is no possibility of understanding why it happened, or of how its recurrence might be prevented.

Aircraft accident investigation is greatly assisted by Accident Data Recorders (ADR[59] – the Black Box) and by Cockpit Voice Recorders (CVR). Modern ADRs record hundreds of parameters of the flight profile and of the engine and system performances. They take 90% of the guesswork out of the investigation. They have been mandatory in civil aviation for decades (in GB since the 1970s). As a result, in competent countries there are few, if any, investigations which produce findings of cause unknown.

The value of ADR and CVR is illustrated by the RAF accident to ZD 576. Two previous BoIs (1987 and 1988) into Chinook fatal accidents had recommended the fitting of ADR and CVR to RAF Chinooks. Seven years later ZD 576 was equipped with neither. For that reason alone it was impossible for the BoI to determine the cause of the accident with objectivity. By failing to fit ADR/CVR the RAF is failing to prevent future similar accidents. It is failing to act with due diligence to protect its crews and their passengers.

Many investigation agencies are charged with making recommendations for accident prevention. As discussed, most are cautious in their pursuit of this objective.

There are few technical difficulties, as we will see in our next chapter. The difficulties lie in persuading investigators that a more proactive approach to prevention will not prejudice their task of investigation. It need not discourage the cooperation of witnesses, because blame is not a part of the prevention process. Prevention addresses factors which can be rectified. It does not address human error in itself, because the fallibility of human beings cannot be eradicated. Prevention addresses the situation which leads to human error. Situations are capable of rectification. Thus the accidents which arise

59 ADR is Accident Data Recorder in RAF parlance, usually Air Data Recorder; now more often Flight Data Recorder or Event Data Recorder. It is the Black Box. Actually it is orange.

from undesirable situations can be prevented. And that is what we want.

We have seen in this chapter that investigations can prevent accident recurrence, that they should, and that they do not. In the next chapter we will see how easily they could fulfil this vital role.

CONCLUSIONS

1 The objectivity of the investigation process must be closely protected by law
2 Closure requires a clear and preferably simple explanation of the cause of an accident
3 The fundamental cause of an accident is always the failure of the system of operation. This is neither clear nor simple
4 The specifics of the failure of the system must be identified, both to facilitate closure and to enable their rectification
5 Responsibility and blame are not the same
6 Blame is irrelevant to prevention
7 Investigation is a tool of prevention
8 The potential of investigation as a tool of prevention is – as yet – widely unrealised

INTRODUCTION TO PART III: THE PREVENTION OF FAILURE

CHAPTERS 8–12

In Part I, Chapters 1–4, we analysed why things go wrong. Our model (to use a posh word) of the world was one in which the mischievous Gods are always seeking to confound our endeavours. Their weapons range from gentle trips via wallops to thunderbolts. Their methods centre on the creative use of three of nature's Laws, the two Murphys and the singular Sod. Additionally, they are ever-watchful for the temptations of easy paths through our system defences; these we offer them by insufficiency in our thinking and by a variety of other self-inflicted injuries. The Gods, being omnipotent, have ruled that the battle be conducted to their laws. This is not fair. And that is the first law of life. Life is not fair. That's the way it is, we cannot change it, so we must get on with it.

In Part II, Chapters 5–7, we sought to learn the lessons of accidents. We reviewed twenty-one cases to see where we went wrong. We found that system failures were spread across the whole system; they were concentrated most heavily in the direction and management functions. We found that, of the three foundations of decision-making, the most common failing was in integrity. We looked at one very thorough investigation and one truly dreadful accident enquiry. We saw that there were impediments to the use of investigations as a tool of accident prevention. We also noted that decision failures and/or self-inflicted injuries were involved in all of the accidents.

In Part III, Chapters 8–12, we now address the prevention of failure.

In Chapter 8 the principle of a new methodology, Failure Path® Analysis (FPA), is explained. FPA formalises modern understanding of system failure. It reveals clearly the falsity, if not falsehood, in convenient and simplistic verdicts of operator error. FPA can be applied both to the investigation of, and to the prevention of system failure.

In Chapters 9–12 we have a careful look at decision-making. The risks to which we are exposed are dictated by our decisions, and unfortunately by the decisions of the hierarchies on whom we are obliged to rely. Sound decision-making is therefore central to the prevention of failure.

CHAPTER 8

FAILURE PATH® ANALYSIS AND ACCIDENT PREVENTION

IN THIS CHAPTER

The prevention of accidents requires the prevention of failure of the system of operation. To achieve this, we must have:

- a method of preventing failures of the system of operation
- a situational context in which the prevention of accidents is a first priority

Failure Path® Analysis (FPA) is a new method of analysing a system of operation. The FPA methodology systematically audits the system of operation; it identifies all the possible opportunities to have averted the accident. This chapter outlines its principles and demonstrates its use.

We have already agreed that accident investigation has (or should have) three purposes. It should establish:

- what happened,
- how it happened, and
- how it can be prevented in the future

In this chapter we apply FPA to three accidents reviewed in Chapters 6 and 7. As an introduction to the methodology, we first apply it to a simple case, the NHS fatality at QMC. Then we apply it to two aircraft accident investigations: US Air Force *Friendly Fire* and RAF Chinook ZD 576. None of these investigations achieved the third purpose, spelling out specifically the steps necessary to the prevention of future similar accidents. The NHS and the Friendly Fire investigations achieved their first two purposes. The RAF Chinook ZD 576 investigation achieved none of its purposes. These applications of FPA show that much of the data needed to prevent the recurrence of a similar accidents was (or should have been) contained in the accident investigation report or study. FPA extracts that data. FPA turns accident investigation into a tool of accident prevention. It enables investigation to fulfil its third, and perhaps its most important, purpose.

Finally we discuss the importance of a context in which the imperative of accident prevention is not submerged by the weight of other imperatives.

FOR SUCCESS, FIRST PREVENT SERIOUS FAILURE

There is one guiding principle which applies to every field of endeavour: avoid serious adverse events. None of us can forecast the future. What we can do is to seek to insulate ourselves against the worst.

Consider your worldly wealth. Most financial journalists are either ignorant or desperate to fill their weekly column or beholden to the investment industry. Most investment advisers make their money by taking a cut of yours, and then by assisting you to lose the rest. They purport to be able to forecast the future. Perhaps in one in ten thousand times they can tell you what will happen but, like statisticians, they damn sure cannot tell you when it will happen. Save time, save money: ignore the lot of them.

Those who are worth their salt are few and far between. You can recognise one if he tells you that he does not understand how the market works, if his fortunes are tied to yours – that is if he does not benefit every time you buy or sell an investment,[60] if he tells you to foreswear all clairvoyance and to stick to a few simple rules, if he tells you that the most important of these rules is to diversify. By thoughtful diversification you

60 And as for those who invest in hedge funds that take a share of any profit you make, whilst not sharing in any loss. Urgent call to the station-master at Milan (Chapter 4, footnote 38).

avoid wipe-out. In other words you prevent an extreme adverse financial event. The essential to keeping in credit is to avoid big deficits.

Consider now the act of choosing a spouse.

Case A: You are in love. Your brains are scrambled, so you are not capable of taking a thoughtful decision.

Case B: If your brains are not scrambled, you are not in love; marriage almost certainly won't work, so stop kidding yourself; don't even think about it.

Going back to Case A.

Case A1: You are totally overcome by love; you are deaf to advice. You are lost. Good luck.

Case A2: Brain still working. How do you take your decision? Read the agony aunts, they have some good insights.[61] They will also tell you to make a list of pros and cons.[62] Ah, you say, soulless – leaves out your feelings. No, it does not; put them on the list; weight them if you think them important. So make the list. Apply the principles of serious adverse event avoidance. Ignore the pros. Concentrate on the cons. If there anything about your putative spouse which in a few years you will find intolerable when sober? Impossible mother, incontinent spender, unpleasant when displeased, useless under pressure, disaster-prone, lazy, thick, humourless, not much fun unclothed?[63] If there is not, then you have a chance of working together as you face the incoming slings and arrows of outrageous etc. If there is, you will have little motivation to persevere in the face of difficulties. The rough will be too rough to take with the smooth. Serious unhappiness will always overwhelm minor happiness. You have your answer. The rule in marriage, as in everything, is to avoid the seriously adverse. Pause to admire the Gods. They have arranged that, at the time that you take one of your life's most important decisions, your brains are scrambled. What a splendid sense of humour.

61 Many have failed relationships behind them, good experience but is it good advertisement for their advice?

62 Military Staff College would tell you to List the Factors Affecting the Choice of Spouse. The Staff College approved answer would be worth reading.

63 Do not take offence, it is your list, not mine.

The starting-point in avoiding the extremely adverse is the careful definition of what constitutes an extreme adverse event. This will depend on the context under consideration. We touched on the need to separate real nasties from trivia in Chapter 4. It is impossible to prevent every single adverse event, no matter how trivial; first, because our world is simply too complex and too dynamic; second, because the endeavour would consume all our resources, leaving none for the chosen activity. Obviously those with tiny brains do not understand this and fixate on trivialities because they cannot get their minds round anything bigger; more about HMG (2008) in Chapter 14. What we need most is a simple distinction between the real nasties and everything else.

You would probably agree that causing the death or injury of a person should be at the top of the list of extreme adverse events. Thus airlines must prioritise the prevention of fatal air accidents, medical authorities of fatal medical accidents, and parents of fatal domestic accidents, etc. This policy of priorities is simple; it is therefore do-able. It can be enshrined in corporate culture and in corporate behaviour, and in human minds and human behaviour.

At the other end of the spectrum, a policy of zero adverse events of any nature, however laudable an ideal, is not do-able. Set as the overriding priority, it brings everything to a grinding halt. Impose on school trips a requirement to rescue within 15 seconds any child who falls into water, then procedures can be put in place. Impose the requirement that no child ever fall into water, then school trips near water cannot take place. School-children will not learn to manage the hazards of water. In consequence some will probably later drown. This will not matter to the twerps responsible for school regulations because it will happen elsewhere and therefore will not affect their targets or their self-esteem.

Having identified the extreme adverse events, the task is then to establish a situational context in which the prevention of such events is the organisational priority. It is not as easy as you might think. We will return to it at the end of this chapter.

THE PREVENTION OF FAILURE

In Chapter 4 we examined the difference between hindsight and foresight. We saw how easy it was to trace the path back from an event, identifying all the steps. We saw that it was impossible to trace a path forward from an insignificant event and to predict where and how it will end. We can see the past. We cannot see the future. End of story.

This is not to denigrate useful methods of failure prediction: HAZOP, Failure Trees and the like. They take the form of a series of speculations based on a 'what if' hypothesis. What-if that pipe burst and that shut-off valve failed? They are laudable planning tools, but they have three fundamental limitations. First, they rapidly become very complicated. Second, when applied to complex systems they are simply not capable of foreseeing all possible interactive failures. Remember Perrow's badly insulated, incomprehensible systems? That is where failure prediction is needed and that is where what-iffing disappears off to iffinity, sorry, infinity. Third, failure prediction is inevitably combined with failure probabilities. The probability of this one-in-a-million failure occurring simultaneously with that one-in-a-million failure is one-in-a-billion. Cannot happen, old boy. Bang. Wrong again. The chances of a take-off coinciding with an engine failure coinciding with a slat failure were calculated at less than one in a billion. This supposed less than one in a billion chance killed 273 at Chicago O'Hare airport on 25 May 1979 (Chapter 5). What did the Nobel Prize winner say (Chapter 3)?

Accident investigation starts with a known event, the accident. It traces the event back through the system of operation, identifying the key failures in the fatal row. Thus accident investigation is a ready-made tool of accident prevention.

PRINCIPLES OF FAILURE PATH® ANALYSIS

A new methodology, Failure Path® Analysis (FPA) promises a real advance in both accident investigation and accident prevention. The details of FPA are protected by patent. A general outline is published here.

FPA is based on the same principle as accident investigation. It starts with a known event and tracks back through the system of operation. It conducts an objective audit of the system of operation. It identifies each and every component in the fatal row. It differs from investigation in that it identifies all the failures, not just the key failures.

Conventionally, investigators work out what happened, and to a certain extent, how it happened. FPA completes the task of establishing how it happened; it then goes on to establish how the accident might have been prevented. FPA adds value to the expertise of investigators. It reshapes the process of investigation into an active tool of prevention.

FPA can be used to prevent accidents before they happen. Again, it starts with a known event, the specific accident which is to be prevented. The process of tracking back identifies all the deficiencies in the system of operation which, if combined into a fatal row, would lead to this accident. These deficiencies can then be remedied. Thus this specific failure of the system of operation and the consequent accident will be prevented.

Founding Concepts of Failure Path® Analysis

Failure Path® Analysis is founded on five concepts, all of which we have already discussed (Chapter 4).

1. The avoidance of extreme adverse events is objective B, one of the three objectives of the system of operation
2. An extreme adverse event constitutes a failure of the system
3. The causes of an extreme adverse event are the failures to prevent it
4. Opportunities to prevent an extreme adverse event always (or nearly always) exist
5. Most of these opportunities are pre-positioned in the system of operation prior to the final event which triggers the accident

The Products of Failure Path® Analysis

Failure Path® Analysis identifies:

1. The Failure Points within the system
2. The Failure Path: the path through the system linking these Failure Points
3. The Locus within the system of each Failure Point
4. The Point of Responsibility for each Failure Point
5. Rectification for each Failure Point

FPA Concept of Causation

We examined (Chapter 4) two concepts of causation.

The scientific concept is that when A follows B as a direct result of B, then B is the cause of A. The legal concept is that when A would not have occurred but for B, then B is a cause of A.

Failure Path® Analysis is rooted in the viewpoint that any operating system must seek to achieve simultaneously the three objectives of conducting the operation, preventing accidents, and conforming to regulations. FPA identifies system failures by asking, not 'what caused A?', but 'what would have prevented A?' Hence the FPA definition that cause is a failure of prevention. It bridges the scientific and legal definitions.

Possible Opportunity to Prevent Loss (PO2PL)

The fundamental unit of FPA is Possible Opportunity to Prevent Loss (PO2PL). This concept is the keystone of the methodology.

A PO2PL is a module within the system which offers an opportunity to prevent an extreme adverse event, or to mitigate its consequences.

- Each PO2PL module constitutes a failure point within the system
- The path linking the PO2PLs is the system failure path
- FPA identifies the locus within the system of each PO2PL

- FPA identifies the point of responsibility for the locus of each PO2PL
- FPA identifies the appropriate rectification for each PO2PL

All will become clear when we apply the methodology to three accidents which we have already discussed. First, the NHS accident at QMC, Nottingham. [64]

FAILURE PATH® ANALYSIS: NHS FATAL ACCIDENT, QMC, NOTTINGHAM, 4 JANUARY 2001

The Event

We discussed this accident briefly in Chapter 6. We focussed on one major failing, the lack of mechanical safeguard to prevent the connection of an intravenous (IV, in a vein) syringe to an intrathecal (IT, in the spine) needle. We return to the event now to analyse it with Failure Path® Analysis. This will determine whether the accident was a single, simple error, or whether it was the product of an extensive failure of the system of operation.

First let us look at the event in more detail.

The 19-year-old male patient was suffering from leukaemia; he was in remission. He was therefore in the maintenance phase of treatment. This entailed his taking oral and weekly drugs at home, and attendance at QMC every 3 months for an IV injection of Vincristine, an IT injection of Cytosine, and oral medication. The patient had missed two of his scheduled appointments in the previous six months, and there were concerns that he may not have been taking his drugs. On 4 January he missed his morning appointment with Dr A at the hospital and did not notify the hospital of his intention to arrive that afternoon.

Dr A qualified as a medical practitioner in Zimbabwe in 1992. He was appointed Locum Consultant Haematologist in August 2000. Dr B took up his appointment at QMC on 2 January 2001. He had been

64 Polite reminder: QMC – Queens Medical Centre, Nottingham.

trained in Syria in 1991 and came to the UK in 1998. He had worked at the Leicester Royal Infirmary in the 11 months previous to moving to QMC. Dr C had joined the staff of QMC in August 2000. He had been on the staff of Ward E17 for five weeks. Neither Dr B nor Dr C had had any kind of formal training to administer cytotoxic chemotherapy. Dr A was the most senior doctor, Dr C the most junior.

Dr A gave a verbal instruction to no one in particular that he wished to see the patient on his arrival. This instruction was not entered in the patient's notes.

The dangers of the IT injection of Vincristine were well-known; they are either fatal or severely damaging. Warnings had been published by the manufacturer, practitioners and the Health Authority. Some previous accidents had been fully reported.

As a safety measure, QMC had instituted a protocol under which the IV and the IT drugs used in cytotoxic chemotherapy were not administered on the same day.

Further protocols were intended to keep IV and IT drugs well apart.

There were a number of similarities in the labelling and in the physical appearance of the IV and IT syringes.

There was no colour difference between the IV Vincristine and the IT Cytosine; both were clear liquids.

The connection on the IV Vincristine syringe fitted the connection on the IT spinal needle.

A code of practice document was available to all staff. The consultant staff of Ward E17, in which the accident occurred, were not aware of its existence. The document was seriously inadequate. Worse, there were two versions in circulation, from one of which there was a major omission.

The induction process for new specialist registrars (a junior grade of doctor) was informal. There was no formal assessment to ascertain that the required standard had been achieved.

Two of the doctors were not born or trained in this country. Questions must arise of whether their medical standards or their command of the English language may have contributed to the confusion, and of whether Dr C was the more reluctant to query Dr B because the latter was foreign. Tricky questions indeed but, for the safety of patients, they must be asked. Pussy footing around political correctness has no place when lives are at stake.

The patient's late arrival short-circuited the hospital's procedures; in consequence, the staff grade doctor experienced in chemotherapy had gone home. The hospital's policy of seeking to accommodate patients' needs put its leaky procedures under further pressure.

Conclusions of Failure Path® Analysis: NHS Accident, QMC, 4 January 2001

There were thirteen broad categories of Possible Opportunities to have Prevented this Loss:

1 *Mechanical Safeguard*: by ensuring that it was not possible to connect any syringe other than an IT syringe to the IT needle
2 *Drug Identification*: by ensuring that all possible steps were taken to distinguish drugs for injection into the spine (intrathecal, IT) from drugs for injection into veins (intravenous, IV). This would include distinctive differences in the shape, size, colour and labelling of their packaging
3 *Drug Handling*: by ensuring that physical separation was maintained between IT drugs and IV drugs
4 *Training*: the training and testing of all new staff to ensure their attainment of the required standards. Further, by ensuring that all staff involved in drug handling and administration were fully aware of the dangerous consequences of the IT injection of an IV drug
5 *Supervision*: by ensuring the necessary supervision of all medical procedures, including the administration of IT injections
6 *Authorisation*: by permitting only authorised staff to conduct critical medical procedures, including the administration of IT injections

7 *Procedures*: by ensuring that administrative procedures reduced to the minimum the possibility of any critical medical accident, including the IT injection of an IV drug into a patient
8 *Protocols*: By ensuring that common practice did not depart from formal procedures in matters of importance
9 *Patient Conduct*: by the conformity of the patient to hospital appointment times
10 *Final Event Handling*: by more informed decisions and actions by the doctors involved
11 *Information Sharing*: by the sharing of information amongst similar units elsewhere to identify best practice. And by a formal system for reporting accidents and incidents
12 *Safety Organisation*: by the establishment within QMC of a safety organisation charged with the pursuit of operational safety
13 *Safety Culture*: by establishing patient safety as the overriding priority for all staff

These thirteen PO2PLs constitute in themselves a massive, near complete crack in the QMC system of operation. A full fracture awaited only a minor final event, in this case a doctor new to the system. There is nothing complicated about any of the PO2PLs. They amount to no more than common-sense. In general terms, all had long been standard practice in aviation.

Easily avoidable accidents such as this will continue in the NHS until its directors wake up to the fact that their safety system is fifty years behind best practice. And take very serious steps to catch up.

Notes on Table 8.1

1 Table 8.1 is an abbreviated version of FPA
2 Column 1 records each Factor. In this case the factors are the broad categories listed directly above
3 Column 2 records the PO2PL, the Possible Opportunity to Prevent Loss which was not taken
4 Column 3 records the Effect Probability. That is the probability with which the taking of the opportunity, the PO2PL, would have

TABLE 8.1 FAILURE PATH® ANALYSIS

NHS ACCIDENT, QMC, 4 JANUARY 2001

FAILURE PATH

Factor	PO2PL	**Effect Probability** Certain	Probable
Mechanical Safeguard	IT conn'n/IT syringe	Certain	
Drug Identification	IT pack colour code		Probable
	IT pack labelling		Probable
	IT pack shape/size		Probable
	IT syringe shape/size		Probable
	IT drug colour code		Probable
Drug Handling	IT/IV drug separation	Certain	
Training	IT awareness		Probable
	Acceptance procedures		Probable
Supervision	Effective Procedure	Certain	
Authorisation	Effective Procedure	Certain	
Procedures	Effective	Certain	
Protocols	Enforce Procedures	Certain	
Patient Conduct	Punctuality	Certain	
Final Event Handling	Procedures	Certain	
	Decisions	Certain	
Information Sharing	Info Share System		Probable
Accident Report System			Probable
Safety Organisation	Safety Audit	Certain	
Safety Culture	Establish	Certain	

AVOIDABLE ACCIDENT

had the effect of preventing the accident. Here, for simplicity, probability is either 'certain' or 'probable'

5 Each of these points is a point of system failure
6 The failure path is the path linking these points. It is the crack in the system, the path through the system defences down which disaster was allowed to travel. A completed failure path is a fatal row

FAILURE PATH® ANALYSIS: SHOOT DOWN OF BLACK HAWKS

We now apply FPA to our summary of Scott Snook's research findings on the shoot down of two US Army Black Hawks by two US Air Force F–15s (Chapter 7). You will remember that the keystone question of FPA is 'Was this an opportunity to have prevented the accident?' Try it for yourself on the event summary in Chapter 7 if you wish, you will find it very straightforward.

Conclusions of Failure Path® Analysis: Shoot Down of Black Hawks

1 We are not attempting an in-depth study of this tragedy. However this Failure Path® Analysis of our summary of the event is sufficient to identify a distressingly large number of PO2PLs. Note the easy identification in Table 8.2 of no less than twelve PO2PLs which would certainly have avoided the tragedy
2 The F–15 pilots and the AWACS crew had opportunities certain to have prevented the tragedy; sadly they failed to use them
3 More important is the context in which the F–15 pilots found themselves, a context set by the US Air Force itself. Some of the cracks had sat in the system uncorrected for over two years. One of them, the inability of units of the US Forces to communicate with each other, appears to be a permanent crack in the US Forces' system. Certainly for years, if not decades, they have not been supplied with compatible radios. In 1983, during the invasion of Grenada, US ground forces

had to use public telephones to talk to the US Navy.[65] In the 1980s the US Navy could not talk to Iranian air traffic control, nor monitor friendly airline frequencies.[66] In 1987 the USS *Vincennes* could not talk to the Iranian Airbus but instead destroyed it. Nor in 1994 could US Air Force fighters talk to the US Army Black Hawk helicopters[67]

4 The morale of the AWACS crew (discussed in Chapter 7) hints of a deep-seated problem. When members of the armed services are fed up to the back teeth, inevitably their commitment and determination will begin to wane. Any member of the AWACS operation could have informed a senior officer in blunt terms of the dangers of the wrong IFF codes and of incompatible radios. Indeed someone may have done so. However, no individual took the system by the scruff of the neck and fixed the problems. If members of the armed forces feel that they are being treated with contempt, then that organisation is rotting on the inside. Not very clever when the organisation's function is to deal out death

65 It was in Grenada that Scott Snook was shot up by the US Air Force and then bombed by the US Navy. Next to him in the mud, also wounded, was the officer specifically assigned to coordinate naval gunfire and fighter support. He was unable to make radio contact with the US ground attack aircraft.

66 Speaking personally, the friendly airlines became progressively less friendly as this situation dragged on.

67 This same US Air Force has a massive nuclear arsenal. Rather worrying.

TABLE 8.2 FAILURE PATH® ANALYSIS

SHOOT DOWN OF BLACK HAWKS, 14 APRIL 1994

Factor	PO2PL	FAILURE PATH Effect Probability Certain	Probable
Major systems failures			
US Air Force/US Army Coord'n	Coordinate	Certain	
Incompatible Radios	Compatible	Certain	
Direction failures			
IFF Code Control	Control	Certain	
AWACS Ops Procedures	Organise	Certain	
AWACS ad hoc Crewing	Address		Probable
AWACS Morale	Address		Probable
F-15 Training	Train	Certain	
AWACS			
Report IFF Codes	Report	Certain	
Record Black Hawks	Record	Certain	
Report Black Hawks to F-15s	Report	Certain	
Control F-15s	Control	Certain	
Initiative	Exercise	Certain	
F-15s			
Closer, Slower, Ident Heli's	+ve Ident	Certain	
Query AWACS: Heli's Ident?	+ve Ident	Certain	
Black Hawks	NO PO2PL		

AVOIDABLE TRAGEDY

FAILURE PATH® ANALYSIS: ACCIDENT TO RAF CHINOOK ZD 576

In our third and final example of FPA, we apply the methodology to the accident to RAF Chinook ZD 576 (Chapter 7).There were many opportunities to have prevented this accident, or to have mitigated its consequences. We will briefly examine the most obvious PO2PLs.

Task

- *The twenty-five VIP Passengers*: Responsible commercial organisations limit the number of key personnel who can travel on the same aircraft. The limit is usually between three and five. The RAF had no limit. It put twenty-five key counter terrorist officers from MI5, RUC, Special Branch, the army, and elsewhere on one aircraft. A limit of five would have saved twenty lives. This is a PO2PL certain to have mitigated the consequences of the accident
- *Fixed Wing Aircraft*: Helicopters are 500 times more dangerous than fixed-wing aircraft. An RAF Hercules could have flown above the weather and the mountains. The use of a Hercules was a PO2PL certain to have prevented the accident
- *Chinook HC1*: The (Mk 2) Chinook HC2 had been rushed into service with a number of uncorrected deficiencies, and against the advice of the RAF's own expert test pilots. The Captain's request for an (Mk 1) HC1 had been refused. An HC1 could have climbed to a safe height above the mountains. The HC2 could not because it might have suffered dangerous icing. The use of a Chinook HC1 is a PO2PL certain to have prevented the accident
- *Stricter Regulation of New Aircraft Release*: Stricter Regulation would have prevented the premature operational use of a new aircraft. It is a PO2PL certain to have prevented the accident
- *Special Forces Crew*: The pilots and crew of the Chinook were special forces. By the nature of their special duties they were habituated to extremely dangerous operations. Further, they did not have the mindset of transport pilots, whose first concern is the safety of their passengers. The task of flying an unsuitable

HC2 in marginal conditions through mountains was clearly dangerous. Any responsible transport pilot would have refused it. Use of a transport crew is a PO2PL certain to have prevented the accident

- *Training Flight*: The operation was categorised as a Training Flight. A regulation forbidding the carriage of civilian passengers on potentially hazardous Training Flights would have prevented this accident. Such a regulation is a PO2PL certain to have prevented the accident

Aircraft Equipment

- *GPWS*: No Ground Proximity Warning System was fitted to the Chinook. It would have given audible and visual warning of imminent collision with the ground (subject to certain technical provisos). GPWS has been mandatory in civil aircraft for decades. A GPWS is a PO2PL which possibly would have prevented the accident
- *Radar*: No radar was fitted to the Chinook. It would have given a clear picture of the approaching coastline. Radar has been mandatory in civil aircraft for decades. Radar is a PO2PL which possibly would have prevented the accident
- *ADR/CVR*: Despite the recommendations of the Boards of Inquiry into two previous Chinook fatal accidents, no ADR/CVR was fitted to the Chinook. ADR/CVR would have revealed the causes of previous accidents and events, and of this accident, and enabled the prevention of their repetition. ADR/CVR is a PO2PL which possibly would have prevented the accident
- *Captain's Authority*: in any responsible civil aircraft operation, the final decision – whether or not to fly – is taken by the Captain. In a military operational setting, the aircraft Captain must follow his orders. The final flight of ZD 576 had no operational urgency nor military necessity whatsoever. The passengers were going to a venue more congenial than N Ireland for a conference and golf. The Captain had not only requested a different aircraft, an HC1, but he had also sought permission to set aside the restrictive HC2 limits if

necessary for reasons of safety. Had RAF Captains the authority to act on their own expert judgement, this flight clearly would not have flown. As stated earlier, a competent transport pilot would certainly not have accepted the flight. The failure to give the Captain total authority over the flight is a PO2PL certain to have prevented the accident

SHF Operational Procedures[68]

- *Instrument Flight Abort (IF Abort)*: this is the emergency procedure to be used by an aircraft which, flying in good weather below its safety height, suddenly finds itself in cloud or other conditions of bad visibility. Close to mountainous terrain, a sudden loss of visual conditions is extremely unpleasant. Pilots are trained to deal with unpleasant conditions without compromising safety. The basis of such training is a well-though-out, standard procedure, learned and tested to a high level of skill. For decades other RAF Commands have had a formal IF abort procedure. The SHF did not. Given the nature of military helicopter operations and the nature of British weather, this is almost unbelievable. The senior officers' explanation, that the appropriate action could be regarded as basic airmanship, is pure codswallop. An overshoot procedure, the drill following an engine failure, and many other situations simpler and less stressful than an IF abort could also be described as basic airmanship. Nonetheless professional pilots have often got them wrong in the heat of the moment. That is why formal procedures are laid down. They are absolutely essential to safe flight operations. A formal IF abort procedure, learned and tested, is a PO2PL which possibly would have prevented the accident
- *Turn-Back Point*: The Chinook was heading for high ground in uncertain – potentially marginal – visibility; this calls for very careful precautions. The most obvious contingency plan would be to fix a turn-back point at a safe distance before the rising ground.

68 SHF is Support Helicopter Force, the RAF unit responsible for the flight.

At this point, if visibility ahead was not good, an immediate left turn would be made on to a safe heading. So why was no turn-back point planned? SHF procedures evidently did not require one. A turn-back point was a PO2PL which possibly would have prevented the accident

- *Pull-Up Point*: The possibility of an unexpected deterioration of visibility after passing the turn-back point should have covered by a final safety precaution, the planning of a pull-up point. This would have been the point at which a maximum rate climb[69] would have taken the Chinook up to safety height before it reached the high ground ahead. SHF procedures evidently did not require a pull-up point to be planned. A pull-up point was a PO2PL which possibly would have prevented the accident
- *VMC Limit*: Visual Meteorological Conditions, the conditions for visual flight, i.e. for looking out of the window to see whether you are the right way up, and where you are going. As opposed to IMC, Instrument Met Conditions, looking at your instruments for the same information.[70] The conditions laid down by the SHF required 1000 m horizontal visibility, 13 seconds flying at 140 knots. 13 seconds is not long but would be safe for skilled pilots flying a serviceable aircraft. However if a serious failure occurred, jammed flight controls or an engine runaway, then 13 seconds would give no margin for safety. Given the known problems of the new Chinook HC2, much better weather conditions for VMC should have been applied to it. Most certainly they should have applied to any flight involving the carriage of passengers. More restrictive conditions for VMC are a PO2PL certain to have prevented the accident

You will note how these marginal VMC conditions combine with the culpable absence of an IF abort procedure to compound the danger to the helicopter and its occupants.

69 A max rate climb is a routine procedure. It is more leisurely than an IF abort, which is an emergency climb.

70 Met is the standard abbreviation for meteorological, easier to say (and easier to spell).

SHF Planning Procedures

- *First Waypoint*: Operational attacks by Special Forces doubtless involve sudden appearances from behind a cliff-face. But, for a passenger transport flight, in forecast conditions of possible poor visibility, it is dangerous beyond belief to set a cliff-face as turning point. The normal, fail-safe procedure would be to offset the waypoint from the cliff by a safe distance, perhaps 5 miles. This would enable the turn on to the next track to be made well before the cliff was reached. A safe offset is a PO2PL certain to have prevented the accident
- *Outbrief*: this is the RAF term for the presentation of the flight plan by the operating crew to an experienced officer for his approval. It should involve the examination of any possible difficulties and the formulation of appropriate contingency plans. Prior to the flight there was no experienced officer available to conduct the outbriefing. This is a deficiency in any circumstances. This flight was probably the RAF's most important of the year in N Ireland, given the value and number of its passengers. In these circumstances the deficiency was grave and culpable. A thorough outbriefing would have covered the choice of the first waypoint and the formulation of contingency plans. It would also have covered the deficiencies in SHF operating procedures (above). A thorough outbriefing was a PO2PL certain to have prevented the accident

Conclusions of Failure Path® Analysis: RAF Chinook ZD 576

1. The start-point of FPA is the accident cause. If, as in this case, the cause cannot be determined with certainty, a full FPA would examine all the most probable causes, tracking back to identify PO2PLs. For example, it would track back from a control malfunction, examining its possible causes and deriving the associated PO2PLs
2. FPA cannot either prove or disprove any of the possible accident causes

3 FPA cannot prove or disprove pilot negligence. However, if we ignore for the moment the fact that the RAF finding of pilot gross negligence is in gross breach of the RAF's own required standard of proof, FPA could track back through the pilots' training and operational records for mention of negligence. Given that negligence in a pilot is a very serious fault, its occurrence would be incompatible with their known high ratings and their selection for special forces. Therefore FPA is likely to find their records exemplary, positive evidence that they were the very opposite of negligent

4 FPA can audit the RAF/SHF system of operation to establish the context in which the pilots found themselves. Our analysis shows that the system was very seriously flawed from beginning to end, starting with the aircraft release for service and ending with the *minutae* of the flight. The system abounded with cracks. The RAF hierarchy was responsible for all of them. The pilots were responsible for none of them. A total fracture, a completed fatal row, was a whisker away. Despatching an unsuitable aircraft in unsuitable weather over an unsuitable route was a very substantial whisker

Notes on Table 8.3

Table 8.3 (below) is the Failure Path® Analysis Summary of the accident to RAF Chinook ZD 576.

1 Column 2 is again the Possible Opportunity to Prevent Loss which was not taken. Taking the first factor as an example, the passengers could have been divided into small groups, each group being transported separately

2 Column 3 is the Effect which taking the PO2PL would have had. In this case, losses may be avoided (A) or reduced (R). TLA is Total Loss Avoided; HLR is Human Loss Reduced; etc. Thus the effect of limiting the number of passengers to five would have reduced the number of lives lost, hence HLR. All other PO2PLs

TABLE 8.3 FAILURE PATH ANALYSIS

ACCIDENT TO RAF CHINOOK ZD 576, 2 JUNE 1994.

FAILURE PATH

Factor	PO2PL	Effect	Effect Probability	
			Certain	Possible
Task				
Passengers, 25 VIPs	Small Groups	HLR	Certain	
A-C/Helicopter	A-C	TLA	Certain	
HC1/HC2	HC1	TLA	Certain	
New A/C Release	Stricter	TLA	Certain	
Spec. Forces Crew	Transport Crew	TLA	Certain	
Training Flt	Stricter Regul'n	TLA	Certain	
Capt. Authority	Empower	TLA	Certain	
A/C Equipment				
GPWS	Fit	TLA		Possible
Radar	Fit	TLA		Possible
ADR/CVR	Fit	TLA		Possible
SHF Operational Procedures				
IF Abort	Formalise	TLA		Possible
Turn Back Pt	Mandate	TLA	Certain	
Pull Up Pt	Mandate	TLA		Possible
VMC Limit	Stricter	TLA	Certain	
SHF Planning Procedures				
Waypoint 1	Offset	TLA	Certain	
Outbrief	Stricter	TLA	Certain	

AVOIDABLE TRAGEDY

would have certainly or possibly have prevented the accident itself, hence TLA

3 Column 4, Probability. The cause of the accident to ZD 576 has not been determined because it was not equipped with ADR/CVR. There were serious flight control and engine malfunctions in the helicopter's record. This suggests that they are the most probable accident scenarios. In this case, the uncertainty about the accident cause underlies the division of the PO2PLs into two groups, Certain or Possible Effect on the prevention of the accident. For instance, if the helicopter suffered a complete and sustained jam of its flight controls, then the fitting of GPWS or Radar might not have prevented the accident; hence 'Possible' effect of PO2PL. Such details are covered in a full Failure Path® Analysis, but are not included in this brief outline of the method.

SUMMARY: THREE DISGRACEFUL ACCIDENTS

All three accidents which we have reviewed in this chapter were easily avoidable. Each was a product of a seriously defective system of operation. We would have hoped that the NHS, the US Air Force, and the RAF were high-reliability organisations. High-reliability is a term used in risk management by those who like to believe that accidents are not normal, but can be prevented by good management. High reliability is not a term which could be applied to any of these three culpably flawed organisations.

The NHS deals in death, supposedly standing between us and the Grim Reaper. The US Air Force and the RAF are killing machines. Further, their tools of trade are themselves extremely dangerous, potential dealers of accidental death to their operators and to anyone in their vicinity. Thus all three of these organisations are highly attuned to the proximity of death. Therefore they should be exemplary in their procedures for preventing inevitable, trivial glitches from triggering an extreme adverse event, and causing accidental death. Far from being exemplary, the failings of their systems of operation were extensive and culpable. Why?

If you have read this book thus far, there is a fair chance that you are persuaded of the view that any simple verdict of operator error is crass. Yet in two of our cases, the QMC and the RAF Chinook accidents, officialdom placed the blame upon the final operator. In the former, the unfortunate doctor failed to use his opportunity to act as crack-stopper of last resort; he was at the end of a culpably flawed system, a shambles which he had joined only two days previously. Similarly the Chinook pilots were on the end of another culpably flawed system, another shambles, and there was no positive evidence whatsoever of their negligence. Only in the US Air Force shoot-down of the Black Hawks did officialdom not place the blame on one of the final operators.

The NHS system at QMC failed in every possible way. Poor procedures were poorly understood, poorly enforced and simply not followed.

The US Air Force openness after the friendly fire tragedy was exemplary. Sad to relate, the US Air Force system of operation was as bad as that of the RAF SHF, but for two different fundamental reasons. The US system failed because of the chronic inability of the US to coordinate its armed forces, and the failure of communications within US Air Force. US inability to coordinate its armed forces has been evident for decades. It underlay the total disaster of Desert One – the attempt to rescue US hostages from Tehran (1980), the 241 fatalities in the bombing of the Marine barracks in Beirut (1983), and the friendly fire fatalities in Grenada (1983). In N Iraq in 1994 the US Army's use of wrong IFF codes and its incompatible radios were accidents which had been waiting to happen for two years. A previous near-miss in which two F.111 fighters had 'popped their safety's and was [sic] getting ready to wax [2 Black Hawks]' came to light in a casual bar-room conversation: 'We were on the trigger when we realized it was you'. Had this identical near-disaster been recorded and disseminated in the official incident reporting system, surely corrective action would have been taken. Vital information was in the informal system; it was never entered in the formal system where it could have been acted upon.

The catastrophic failure of the operational system of the RAF Support Helicopter Force system occurred for two fundamental reasons. First, it was certainly not fit for the purpose of the non-operational transportation of passengers. It was probably not fit for any purpose. Second, the RAF is a military organisation. Dependent on the style of the individual unit, and on the character of its senior officers, the questioning of orders may well be a court martial offence. The direction of the unit involved, the SHF, appears to have been rigid, unthinking and authoritarian. The captain of the Chinook had indeed gently questioned his orders; he was threatened with disciplinary action.

In all three cases, complacency can be identified. In the NHS and the US Air Force cases, procedures had been flawed for an extended period and no accident had resulted. Alertness to the possibility of failure had therefore diminished. In the case of the RAF Chinook, the complacency was the amazing assumption that a new helicopter, hastened into service with known unreliability, was safe to undertake a task to which it was totally unsuited.

Further, in the cases of the NHS and of the US Air Force the sands had shifted. Practice had departed from procedure.

THE PREVENTION OF FAILURE

Each of these tragedies could have been prevented had the daily operational imperative (Task A) not totally submerged the safety imperative (Task B). Obviously in organisations such as these, the safety imperative cannot be given absolute precedence over the operational imperative (not that in the Chinook case there was an operational imperative). Nonetheless the safety imperative must always be in mind. To forget it is to invite self-inflicted injury. And the NHS, the US Air Force and the RAF forgot it. The injury was not to those who forgot; it was not to officialdom. It was to those who had entrusted themselves to itscare.

If such tragedies are to be prevented, the status of the safety imperative must be reasserted. This requires a four-point programme.

1 First and foremost, those in charge must be given responsibility for accident prevention. They must be made aware that if there are culpable accidents on their watch, their careers will be forfeit. This will ensure their close and sincere interest in accident prevention
2 Control of accident investigation must be removed from the management chain and placed in the hands of an independent expert body
3 A dedicated safety inspectorate must be established within such organisations and empowered to pursue safety. Their primary task must be:
4 The establishment of a safety culture

CONCLUSIONS

1 The investigation of each of the three accidents failed in its third and arguably most important purpose, accident prevention. None achieved a result which could actively prevent the recurrence of similar accidents
2 In all three cases, the application of FPA turns investigation into a tool of accident prevention. Failure Path® Analysis thus enables investigation to achieve this most important purpose
3 Each of these tragedies would have been prevented by a specialised safety unit using Failure Path® Analysis or a similar audit system
4 The safeguarding of the safety imperative requires the placing of responsibility for safety on the individuals at the highest echelons of the organisations
5 The safeguarding of the safety imperative requires organisations to maintain a dedicated safety unit, tasked to conduct a rolling safety audit of the system of operation
6 The safety unit should be in permanent pursuit of the safety imperative

7 The safety unit should most certainly not take responsibility for safety out of the hands of the operators. On the contrary, it should tighten their grip upon it
8 The safety unit should set a situational context in which all personnel with opportunities to prevent accidents will take those opportunities
9 The head of the safety unit should have no respect for authority, no career prospects, substantial independent means, and a peerage – or no wish for one

CHAPTER 9

THE ANATOMY OF A DECISION

WHAT IS A DECISION?

We will use the word 'decision' only to relate to action to be taken in the future. Let us see why.

It is our actions, not our thoughts, which take us into or away from danger. We can have silly thoughts without endangering ourselves, whereas silly actions may have serious outcomes.[71] Therefore we use 'decision' only in reference to action.

And why the *future*? Because our present or future actions have no impact on the past. You might indeed 'decide' that last night's escapade was a mistake, whether it was the seventh glass of bad Bulgarian red, the unfortunate gesture to the policeman, the regrettable one night stand, or indeed the whole evening. But decide all you like, there is no action which you can take now to change the facts of yesterday. This morning you can only *realise* what you did; you cannot undo it. 'I didn't!' 'Afraid you did.' Tough. But don't worry. We have all been there. [72]

SHOULD WE CARE ABOUT DECISION-MAKING?

We should care, and very much, because it is decision-making that determines our exposure to risk.

71 To *think* that front-row forwards are fat boys is safe. To *act* by so informing them would not be safe.

72 The Moving Finger Writes; and, having writ,
Moves On: nor all thy Piety nor Wit
Shall Lure It back to cancel half a Line,
Nor all thy Tears wash out a Word of it.

Thought Safe, Action Dangerous

Some of that decision-making is in our own hands. Unfortunately much of it is in the hands of the hierarchies which control most aspects of our lives. Nonetheless, the better we understand decision-making, the better will be our own decisions. Equally important, the better will we be able to assess the competence of the hierarchies. You will find that assessment disquieting; the hierarchies will find it deeply displeasing (Chapter 14).

OBJECTIVE AND SUBJECTIVE QUESTIONS

Let us carefully distinguish between objective and subjective questions.

We start by defining *facts* as *objective.*

By this definition, any simple question of fact will have an objective answer. What is 2 and 2? Mathematics provides the objective answer. What is the French for aeroplane? The dictionary provides the objective answer. Some questions may be extremely complicated; advanced mathematics, hieroglyphics, astronomy or whatever. They remain questions; there is an objective answer to be found.

This is true of more complicated questions of fact. Let us use the term *criterion* for the *objective basis* of an answer based on knowledge/fact.

Am I taller than my son? The answer to this question has the objective criterion: who has more inches of height? Objective answer: Malik junior. He has seventy-two, dwarfing Malik senior at sixty-nine, and pleased about it. Easy.

But, is Malik senior a better man than Gunga Din? No. Based on what criterion? Based on Mr Kipling's judgement. Sorry, I won't accept that, Mr Kipling has never met me; base it on something objective; tell him to push off and bake cakes. Right, we will base it on virtue. And what is virtue based on? Well, Socrates says it is this, Aristotle says it is that, St Thomas Aquinas says it is something else. In other words, no one has succeeded in identifying the irreducible essence of virtue, the incontestable unit in which to measure it. How then can we decide whether I am less virtuous than Gunga Din? There is no objective criterion. It must be a matter of judgement. We must find a *subjective* basis on which to make that judgement.

OUR TERMINOLOGY

It is of fundamental importance to distinguish an *objective* criterion from a *subjective* premise, and an *objective* answer from *subjective* decision. To make this distinction we will use the following terminology:

- *Facts* are objective.
- *Criteria* are matters of fact. Both are objective.
- *Answers* relate to facts and matters of fact. All are objective.
- *Questions* have answers. Both are objective.
- *Premises* are judgements. Both are subjective.
- *Decisions* are based on premises. Both are subjective.
- *Problems* are resolved by decisions. Both are subjective.

You will very soon find these distinctions helpful. Faithful promise. The Table below gives you a picture of them.

TABLE 9.1 OBJECTIVE VERSUS SUBJECTIVE

Objective	*Subjective*
Facts	Problems of Facts in different units
Questions of Facts in same units	Answers to Problems of Facts in different units
Answers to Questions of Facts in same units	
Criteria	Problems(Choices) between Criteria in different units
	Decisions on Criteria in different units
	Qualities
	Problems of Quality
	Problems of Comparison of Quality
	Premises (Bases of Decisions)
	Decisions

That is all very trite and obvious, you say, and generously award the author an MBO.[73]

But am I wasting your time? I hope not. Case Study 9.1.i (p 201) shows that we do not take the decision when we 'decide' on the car. We take the decision *when we decide the criterion.* Important. After we have decided our criterion, it is all plain sailing; all that remains is an objective question of fact to which there will be an objective answer.

THE IMPERATIVE

Just a minute. We are discussing the outcome of our decision-making task. What about its beginning? What imperative or objective started the process in the first place?

In our present example, our imperative might be anything from serious need (demise of last car) to simple self-indulgence (bling for a poseur). The imperative may or may not be the same as the objective. Either way it is the starting-point, the keystone of the decision-making process. Choose the wrong imperative, and we put our ladder up the wrong tree. Therefore we must think hard about it.

Note that in formal situations, the organisation's imperative (what decision is best for the organisation?) may be very different to the imperative which a decision-maker brings to the table (what decision is best for me?), or to the imperative of safety (what is best for our passengers?). Such conflicts often prompt the failure of integrity. We saw in Chapters 5 and 6, that of the three essentials in the decision-making process, the most common failure was of integrity. We now know why.

The imperative is the most important of all premises. It is the founding premise for the whole decision-making process.

PREMISES, PREMISES

We elected to choose the fastest car, Case Study 9.1.i, p 201.

Our subjective judgement was that speed was more important than comfort or cost.

73 Master of the Bleeding Obvious, Fawlty University (which we will not abbreviate).

CASE STUDY 9.1.i
DECISION TIME: WHICH CAR?

We have to decide which car to buy. We have three Lists.

- List A is of the fastest cars
- List B is of the cars with lumbar support
- List C is of the cheapest cars

How do we decide from which List we should select our car? The three criteria – speed, lumbar support, or price – are themselves objective. But they are in different units. Deciding between different units is subjective. We need a premise on which to base this decision.

If our name were Clarkson, our premise would be that speed is preferable to comfort or economy. We would decide to choose from List A.

If our back were playing up, our premise would be that lumbar support is preferable to speed or economy. We would decide to choose from List B.

If our bank manager were playing up, our premise would be that price is preferable to speed or comfort. We would decide to choose from List C.

When we subjectively decide our premise, we thereby decide our criterion: speed, comfort, or economy. Thus, although the criterion is itself objective, please make careful note that the *choice* of the criterion is *subjective.*

Let us buy the fastest before Clarkson wrecks it on the test-track.

Which is the fastest car? This is not a subjective problem, needing a decision. It is an objective question of fact. It has an objective answer, to be found in the cars' performance data. We consult List A.

But upon what basis did we make this judgement?

It was on the basis that speed makes us feel good. And why does it make us feel good? Because we enjoy speed and because we like to show off. Why do we enjoy speed? Because we enjoy the feeling of speed. Why do we like to show off? Because we enjoy the feeling of admiring glances. And why do we enjoy those feelings? Because they give pleasurable physiological and psychological sensations. Hormones whizz about and the pleasure centres of the brain light up.

Let us stop there; we have arrived at feelings. It does not matter to us that we have not arrived at the ultimate, fundamental premise, the basis of our thought processes. Seriously clever philosophers have been seeking this irreducible essence for millennia and they have not found it. Perhaps they never will; it may be beyond our mental capabilities. If we try to chase down the chain of premises to the very last premise, we are likely to disappear into increasing confusion, if not up our fundamental premises. Let us not do that.

CASE STUDY 9.1.ii COMPONENTS OF THE DECISION: WHICH CAR?

START HERE

Subjective Components	
Imperative	*We must purchase a new car.*
Beliefs and Values	(See next paragraph)
Premise	*More Good Feelings for us from Speed and Attention* than from Comfort or Parsimony.
Judgement	Decision is that our Criterion will be Speed
Objective Components	
Criterion: Speed	Therefore *choose from List A.*
Data	*Consult* road test *Data.* Which of Cars A is fastest?

Choice (objective)	Car A 3, Batmobile is fastest. *Place Order.*
END HERE	(Drive Carefully)

BELIEFS

We have just seen that some premises are based directly on feelings. Others are based on beliefs. Beliefs are not facts to be established objectively by reference to textbooks, dictionaries or encyclopaedias. Beliefs are matters of opinion or faith; they are feelings. Thus all premises are based on feelings.

Arguing matters of opinion is tricky. A believes it. B does not. By definition there is no objective evidence, so what kind of evidence is there to deploy in debate? I believe this versus you believe that. Who wins? I believe that I did. You believe that you did. Impasse.

We have pleasurable feelings when our beliefs are confirmed. If someone disagrees with you and tells you how wrong you are, you will not feel very pleased. If someone agrees with you and tells you how right you are, you will feel a warm glow. If you cannot remember when last you felt that warm glow, it is high time you found better company.

Thus it appears that all premises are ultimately based on feelings, delight at a warm glow, dislike of unpleasant sensation. This means that all premises are subjective. It means that all decisions have a subjective basis.

BELIEFS ARE VALUES

Our beliefs are our values. These subjective feelings are the most fundamental guidelines of our lives. Life continually impacts upon them. Conflict between the values is more likely than not. How much money would Mr Snitch need (self-interest value) to give evidence against his friends (personal value)? To steal from a pensioner (moral value)? Would you have an affair (self-interest value) with your irresistibly sexy, married neighbour (moral value)? In any specific case, some value or value-mix will emerge as the winner of this conflict. This winner will be that value/value-mix which for that individual, in that situation, best meets his/her current dominant need, and best expresses his/her dominant feeling.

Values underlie the personal premises upon which each individual makes the decisions which guide their life. Self-restraint (a moral value) may go out of the same door as you if the neighbour is irresistibly sexy (self-interest value). The extreme case is the individual who voluntarily sacrifices their own life to save someone else's. Such individuals choose to lose their life rather than to lose their self-esteem. One self-interest value is subordinated to another.

It seems that in seeking the fundamental premises of decision-making, we can go no further back than feelings based on values. Malik's First Law of Life states that *all people and organisations will always pursue their own interests, as they see them at the time.*[74] Their interests are based on their values. This underlines the role of values as the keystone of society. A society cannot work harmoniously and remain cohesive unless its members share the same core values.

MY PREMISES ARE MY FEELINGS; TAKE CARE

Some of our attachment to our values may be provided by objective knowledge, including the knowledge of circumstances where they provide sound guidance. But fundamentally we are attached to our values by our subjective feelings.

The depth of feeling which attaches to a value is the measure of its importance to us. In general it is likely to be surprisingly deep. Challenges to values, even when trivial, can be incendiary. Weird hair and clothes (social values) can bring on apoplexy. Criticism of the current vogue in popular music (social value) enrages its adherents. Resolution of differences is not easy. There are no objective criteria on which to base a determination. This gives immense practical difficulties when deeply held beliefs must be challenged. Seek to resolve the issues of, for example, multi-culturalism, abortion, or the environment. You will wait in vain for the disputants to agree premises upon which to base a reasoned discussion, because any such premises may implicitly throw doubt on their beliefs. Warning: Do not try this at home.

74 Malik's Laws of Life are listed in the Appendix to Chapter 15.

Then consider – from a safe distance – the difficulties of finding a basis on which to resolve the differences between Israel and the Palestinians, or between different religions, or between different sects of the same religion. As an aside, we may note that serious fundamental differences may themselves be impossible to resolve, because no premises can ever be agreed. Nonetheless, such conflicts can be brought to an end by persuading the warring parties to agree to differ peacefully.

CASE STUDY 9.2 REACHING AGREEMENT

John Adams tells the following anecdote in his highly readable book, *Risk.* Over 150 years ago the Reverend Sydney Smith was being taken on a conducted tour of an Edinburgh slum. Down a narrow alley between two high-rise buildings he came upon two women shrieking abuse at each other across the alley. Smith stopped, looked up, and listened. He then shook his head and walked on, lamenting, 'they'll never agree; they're arguing from different premises.'

CASE STUDY 9.3 OUCH, YOU ARE TREADING ON MY VALUES

At an informal, off-the-record gathering in 2005, Larry Summers, President of Harvard University, was discussing why there were fewer women than men at the highest echelons of science and mathematics. He postulated that there might be innate differences in the abilities of men and women, i.e. that men may be better at some things and women better at others. This would strike many of us as plain common-sense. Note that he was not propounding that women were an inferior sex, a viewpoint confined to the Taliban and their intellectual equals. But, O Boy, did Larry Summers tread on a contemporary value.

Professor Nancy Hopkins, feminist activist and MIT biologist, had a marvellous hissy fit: '"I felt I was going to be sick. My heart was pounding and my breath was shallow. I was extremely upset.' She fled the room before Summers finished speaking (and rushed off to tell the newspapers) about her wrenching experience. Professor Camille Puglia, whom few would call a shrinking violet, put the incident in perspective. In Professor Puglia's words (reordered), 'to question the dominant dogma even an iota brings the thought police out in shrieking mobs to your door.'

A hissy fit and a flounce do not constitute a viable academic methodology. Academics are, or should be, engaged in the pursuit of objective truth. Such truth is founded on its underlying premises. Therefore the search for truth must start with an examination of those premises. Those who cannot stand the heat of that search should get back in the kitchen.

Some years ago I attended classes in philosophy at the Marlborough College Summer School. The resident philosopher, Anthony Reynell, advanced the premise of Bishop Berkeley. The good bishop, he said, argued that there was no reality outside the mind. I disagreed politely. 'What a lot of crap.' 'Oh,' said Anthony, 'yours is a most interesting viewpoint, Omar.'

Anthony was a true philosopher. He did not mind his premises broken into in the search for truth.

Compare and contrast, as they say.

In case you were hoping that Nancy Hopkins was a one-off, you should consider the more recent case of Priya Venkatesan. In May 2008 this Ivy League professor, who lectured on 'French narrative theory' and 'eco-feminism', complained that students questioned her knowledge. When a student, to applause, produced evidence challenging her theory, she denounced their 'fascist demagoguery' and began preparing a lawsuit. Polite words fail me.

THE SUCCESS OF A DECISION

We have distinguished between objective answer and subjective decision. Answer is based upon data. Decision is based on feeling. Armed with this finding, we can now examine our real interest, the success of our decisions. What, we must ask, do we mean by a successful decision?

DECIDE TODAY, PAY TOMORROW

We know that underlying every decision is the mental premise that it will yield good feelings. We were faced with the task of buying a car. We decided that speed would give us better feelings than comfort or economy. This premise yielded our criterion – speed – on which to base our selection.

But, better feelings when? Now, at the time of the decision? Or at some time in the future?

Present bad feelings may be accepted in the interests of future long-term good feelings, (present crash diet against future unobstructed view of toes).

Equally, the opposite may be the case. Present good feelings may be replaced by future bad feelings (present extravagant purchases against future crippling debt).[75]

The point is that *it is the feelings present at the time of the decision that determine the decision*. The present moment is king. But the present moment is very soon replaced by the future. Any failure to consider future feelings is an almost certain self-inflicted injury.

The feelings which matter are those when, at some later time or times, the success of the decision is decided. *It is future feelings that determine the success, or otherwise, of the decision.*

At whatever time in the future we choose to judge the success of the outcome, several conditions must be met. Let us revisit our case study Which Car? to identify them.

This case study shows that the conditions when we take a decision probably will not be the same as the conditions when, at some later time, we judge its success. It shows that we decide today, but pay tomorrow.

75 Let us not make reference to the contents of a lady's wardrobe. Present jest in haste, against future repentance at leisure, at length, and probably at considerable cost.

This does nothing to make our decisions easier. But it does add to the fun of life.

CASE STUDY 9.1.iii
CONDITIONS FOR THE SUCCESS OF OUR DECISION: WHICH CAR?

Our Choice must meet its specification or our expectations.
– The Batmobile must achieve its expected performance.

Our Data must have been correct.
– The performance figures of all cars on the list must have been accurate.

The Criterion must still be valid.
– Speed must give us the expected good feelings. These good feelings, our criterion, must not be overwhelmed by bad feelings arising from the bills from the wife's coiffeuse, from the Batmobile's unreliability, from the unwelcome attention of the traffic police, etc.

The Premise must still be valid.
– We must still value the good feelings of speed above the good feelings of comfort, or of money in the bank, or of sundry other alternatives.

The Imperative must still be valid.
– The imperative was that we needed to purchase a new car. This imperative might be invalid if, for example, we now worked from home or if we had lost our licence.

The World must not have changed and thereby invalidated our decision.
– The world which now prevails must not be detrimental to, or invalidate, our decision. The idiot local council must not have put Batmobile-scraping speed bumps on our road. The need must not have arisen to transport a small addition to

the family and a large perambulator, or our elderly Mum and a wheelchair. We must not have been posted abroad. New Labour must not be seeking to refinance its managerial incompetence or its re-election prospects by raising the Road Fund Tax on Batmobiles to £largemillions.[76]

SUMMARY: THE ANATOMY OF A DECISION:

1. We must distinguish between objective questions and subjective problems
2. Objective questions have objective answers
3. An objective answer, if not a simple matter of fact, is based on an objective criterion
4. The choice between objective criteria is subjective
5. Subjective problems relate either to qualities (which cannot be measured in objective units) or to criteria/quantities which are measured in different units
6. Subjective problems have subjective resolutions
7. Subjective resolutions are based on premises
8. Premises are based on beliefs and values
9. Beliefs and values are subjective
10. Therefore premises are subjective
11. A decision is a subjective choice of premises
12. Decisions are based on feelings at the time of the decision
13. The success of a decision is assessed at a later time

DECISION: CONCLUSION

All decisions are subjective. Man may be a rational animal but, in final analysis, he is ruled by his feelings.[77]

76 Modesty would normally have prevented me from drawing your attention to the prescience of these words. However it is too good an example of a topic to which we will shortly come, the fact that past performance may be an excellent guide to future performance (Chapter 12).

77 *Pace* Aristotle's feelings.

CHAPTER 10

DECISION-MAKING: THE WORLD

NAVIGATING THE WORLD

There are three separate inputs into the decision-making process:

- The *External Situation*, the factors from the outside world which are pertinent to our problem
- The *Internal Situation*, factors from the mind which bear upon the problem and upon the decision-making process
- The *Decision-Making Method/s* which we chose to utilise

The better that we understand these inputs, the better we can master the decision-making process.

To get a feel for the decision-making process, picture yourself about to go for a drive, but with a difference. You may choose your destination. You are not sure about the intervening terrain; you have a navigation system which is not entirely reliable; you have a choice of vehicle.

Your chosen destination is the *objective* of your decision.

The terrain is the *external situation*, the world through which you must navigate. It is the subject of this chapter.

The navigation system is your *mind*; it must guide your decision. The mind is the subject of Chapter 11.

The vehicle is the *decision-making method* which you select to convey you, unless you elect to go on foot. Decision-making methods are discussed in Chapter 12, but briefly, so don't worry.

Our study produces a new theory of decision-making, the Dolly Mixture Selection Box Theory, clearer, better and lots more useful than the competition (Chapter 12).

In this chapter we examine the external situation. We live in a vast and complex world, and most of its factors are irrelevant to any single decision-making task. But some are relevant, some important, and some are crucial; the last we call killer factors. To overlook any of these is to assure a self-inflicted injury.

THEORIES OF DECISION-MAKING

First let us look at the evolution of the theories of decision-making.

Whether the human mind has improved over the course of history is seriously to be doubted. We do know that for over 2000 years theorists held a somewhat optimistic view of human decision-making. In recent times a much more pessimistic view has established ascendancy.

Aristotle (350 BC) famously stated that man was a rational animal. He certainly was a genius. He may perhaps have been close to a totally rational man. The fact is that in 2300 years and umpteen generations, there has not been another Aristotle. Therefore it is fair to say that there is good evidence against the general validity of his theory. Man is indeed capable of reasoning, that is of rationality. He (and most certainly she) is also capable of being damned unreasonable, so the proposition that Man is always reasonable, full stop, falls at the first fence. Capable of reason, yes; always reasonable, you must be joking.

Later, European philosophers expressed their philosophers' view that, by following the laws of philosophy, man could be a dispassionate, coldly logical thinker.

Jeremy Bentham (1770) took the important first step away from theories of dispassionate decision-making, propounding his Utilitarian theory that man simply pursued his own best interests. So, hurrah for Jeremy Bentham.

Modern economists broadly agreed, expressing their economists' view that man pursued his economic best interests (they would, wouldn't they?). They took us back to man as a machine, a calculator which computed the Subjective Expected Utility of each option in order to determine, and thus to decide upon, the best. Economic rationality cannot handle multiple criteria with objectivity; its apparent accuracy is often plain wrong. Consult your own experience. Economic rationality may indeed be amongst your considerations, but you are certainly not ruled by it, even if by misfortune you are an economist. Cold, clunk-click calculation every trip, I think not. This model of the mind as a cold, economic, calculating machine marks the peak of the aren't-we-clever theories of decision-making. The economists are still at it. Now they are claiming that common-sense and rationality in decision-making are methodologies of economics. So, boo to the economists.

The first boot into this self-satisfaction was the Theory of Bounded Rationality (1957). This theory, attributed to Herbert Simon, says that we are rational only within certain bounds. The realities of the world, our temperaments, the nature of our minds, the pressures of time, etc., put limits upon our rationality.

Then came the big fall. The splendidly named Garbage Can Theory (1972) attributed to Herbert Simon (again) and colleagues, was a well-aimed, destructive missile into smug ideas of man as a smooth decision-making machine. This theory holds that decision-making is a lucky dip, a process of rummaging through the garbage-can of ready-made solutions. It observes that in organisations, solutions are looking for problems and decision-makers are looking after themselves. Not all the organisation's interests are represented at every meeting. The resulting decision is unlikely to be the one that meets the needs of the organisation. Each of those in attendance will advocate the solution which best suits their own current needs, probably what their boss wants or, if the boss is not looking, what the attendee personally wants. Thus the decision is extracted from a garbage can of ready-made solutions. The surgeon's ready-made solution to a health problem is surgery, the pharmacist's is drugs, the psychologist's psychology, etc., ad nauseam, until you ask the sergeant-major. He will give you the sergeant-major's ready-made solution. Get your expletive deleted act together. And get your same expletive hair cut.

The aren't-we-clever theories which view the mind as a processor – data goes in, decision comes out – is reasonable as a physical picture of the process. As a working model, it is seriously wrong. The mind is indeed the processor, but it also makes a major input into the process. Unfortunately this input may fatally mislead the decision. We ignore it at our peril.

FACTORS OF THE WORLD, CONTROLLED AND UNCONTROLLED

The *external situation* is the terrain across which we must find our way if we are to arrive at our destination, that is to reach our decision. The terrain is the battlefield on which the success of our decision is determined. Obviously if we fail to identify its key features: impassable mountains, raging rivers, swamps, etc., we will never arrive at our destination. And let's not forget the other perils; predators – our enemies and competitors, regulators of all shapes and sizes – traffic wardens and other undesirables, changing conditions – the weather, and the mischief-making Gods. There are countless millions of factors which might affect our journey. Compiling a list of them would keep our Mongol hordes of government nannies happy in their useless, comfortable, index-linked pensioned lives until the end of time. And just imagine if they brought chaos theory into their calculations. 'Don't kick that ball, Jonny. You might knock a gentleman off his bike in Caracas.' [78]

We need to identify which of the world's innumerable factors will make an appearance on our route. These, the active factors, determine the success of our decision. They define its context.

It will greatly assist our decision-making if we distinguish the factors under our control (*controlled factors*) from those not under our control (*uncontrolled factors).*

78 And reveal Australians to be cry-babies. (Subtle rugby joke). Kick it, Jonny.

To see the importance of this distinction, we will take a very simple business school case study.

It is a gripping tale of tile manufacture. Try to control your excitement.

THEREBY HANGS A TILE

First we need to identify the factors which are key to our success. We apply our conditions for the success of the decision; these we identified in Case Study 9.1.iii.

Case Study 10.1.i sets the scene.

Case Study 10.1.ii identifies the key factors.

Case Study 10.1.iii distinguishes the controlled factors (Controlled), from the uncontrolled factors (Uncontrolled).

CASE STUDY 10.1.i
THEREBY HANGS A TILE

We are a tile factory. We are efficient. We make good tiles. But we do not make a profit. We are in danger of going out of business. We decide that a review of our product range is the key to our return to profitability. Should we focus on a specialist niche with a small range of high quality tiles? Or should we make a wide range of inexpensive tiles? Discuss.

We will make it easy. Here we go.

We first apply the model of decision-making (Case Study 9.1.ii) which we developed in Chapter 9.

Imperative Return to profitability within one year.

Premise We judge that, if we chose our new range correctly, sales will be lucrative and we will achieve the Imperative.

Criteria We decide that superiority to our competitors' products in *design* and in *price* (lower) are our two criteria.

Facts Data on the design and price of each of our two possible ranges is provided by our production department (Fred). This data is compared with the data on our competitors' products by our marketing department (Tracy).

Choice We have two possible choices: small, specialist range of high quality tiles or, wide range of inexpensive tiles. The data provided by our marketing department shows that the specialist range would be the more profitable. We choose it as our new product range.

Next, we identify the key factors in the external situation.

CASE STUDY 10.1.ii SPOT THE KEY FACTORS (they are in *italics*)

The Choice must, in its performance, meet criteria. Our specialist range of tiles must achieve *superior design* and *lower price.*

The Facts must be correct. Our *data on design* and *on price*, and *on our competitors' products* supplied by Fred (production) and Tracy (marketing), must be accurate.

The Criteria must be valid. Our criteria were *design and price.* It might transpire that we were unable to manufacture the planned quantities because the required high quality of our specialist range caused production difficulties. In which case, our criteria would be invalidated by our omission of *ease of production.* Or our competitors, the rotters, might have improved their *designs* and lowered their *prices.*

The Premise must be valid. Our *premise* was that lucrative sales of our new range would return us to profitability. But what if we are manufacturing on demand and offering four weeks delivery, whereas our well-funded competitors can afford to stockpile and offer *overnight delivery*? Or what if tiles went *right out of fashion*? Either could invalidate our premise. Ouch.

The Imperative must still be valid. Our *imperative* was a return to profitability within one year. What if we had *five years* in which to return to profitability? With perhaps a Chancellor who was a sentient human being, not a glowering Machanical M'Doom[79] intent on funding his MacDogma by squeezing us of all our vital forces. What if he offered generous tax relief to fund modernisation and the return to sound, long-term profitability? I know, I know, totally unrealistic, but it is only an example. There are plenty of realistic scenarios which could invalidate our imperative, eg. generous refinancing, or foreclosure by the bank of loans. In all such cases it would be, 'Back to the drawing-board, Hoskins'.[80]

The Controlled and Uncontrolled Factors. We can see that, in this case, some elements of facts, criteria, premise, and imperative are controlled factors, under our control. But a most important element, their future validity, is beset with uncontrolled factors, outside our control.

79 Of the same clan as Kipling's Mechanical M'Andrew.

80 Traditional aviation observation when, due to false design assumptions, an aircraft falls apart on its first flight.

Next we distinguish the *Controlled Factors* from the *Uncontrolled Factors.*

CASE STUDY 10.1.iii CONTROLLED AND UNCONTROLLED FACTORS

Our Criteria are:
– Design, (Controlled)
– Price, (Controlled).

Our Criteria could be invalidated by:
– Omission of ease of production, (Controlled)
– Improvement of designs by our competitors, (Uncontrolled).
– Lowering of prices by our competitors, (Uncontrolled).

Our Facts could be invalidated by:
– Defective data from our production or marketing departments, (Controlled).

Our Choice could be invalidated by
– Our Failure to improve design, (Controlled)
– Our Failure to lower price, (Controlled).

Our Premise (lucrative sales) could be invalidated by
– Our well-funded competitors offering overnight delivery, (Uncontrolled).
– Tiles going out of fashion, (Uncontrolled).

Our Imperative (return to profitability within one year) could be invalidated by:
– Chancellor glowering Machanical M'Doom being replaced by a sentient human being, who gives us five years to return to profitability, and

– Generous tax relief, (Uncontrolled)
– Generous refinancing (Uncontrolled), or
– Foreclosure by the bank of loans, (Uncontrolled).

And, please no complacency. Let us not forget:

– Unforeseen Factors (Uncontrolled and/or Controlled).

We would hope that we had given consideration to all of these important factors in our decision-making. We see that only four: design, price, ease of production and data, are controlled. Most are deep in the uncontrolled situation, outside our control. For these we should have prepared contingency plans. It is a tautology that the *specific form* of the unexpected cannot be foreseen. But the *possibility* of the unexpected can be foreseen. We can, and should, build enough general slack into the system (reserve funds, stocks or other resources), to cope with an unforeseen factor, a shock in the short term. This will ensure that the shock does not instantly destroy us. It will enable us to survive for long enough to generate an appropriate, specific, response.

UNCONTROLLED FACTORS RULE. DAMN. NOT OK

Case Study 10.1 shows that the controlled factors and uncontrolled factors are interwoven inextricably into our decision.

However well we control our controlled factors, the success of our decision may be greatly, even totally, dependent on the factors beyond our control, the uncontrolled factors. They have enormous power. On the one hand, they can give us success whatever we decide (e.g. massive increase in demand for tiles). On the other hand, and more likely, they can deny us success. The uncontrolled factors have us by the vitals. We must give them our closest attention.

Our decision succeeds or fails on the battlefield where the controlled and uncontrolled factors meet. Here nasty predators, our competitors, roam. Glowering M'Doom maclooms. And the Gods come out to play. Murphy, well-meaning but all thumbs, is ever-ready to lend his disastrous hand. Sod, dirty trickster *par excellence*, is spring-loaded for instant action. Thunderbolts are sniff-triggered to the faintest whiff of hubris. The pitfalls of self-inflicted injuries abound; buts are built in; agents double; sands shift. Here the Gods Rule. If they approve our battle performance, our risk management, they may permit our endeavours to succeed. If not, see footnote 80.

UNCONTROLLED SITUATIONS – EASY AND DIFFICULT

Let us have a closer look at the uncontrolled situation, that large and important part of the context of our decision which is not under our control.

An uncontrolled situation which is simple and static poses us few problems. Because it is simple we can comprehend it. Because it is static, it will not undergo changes which could invalidate our decision.

An uncontrolled situation which is complex and dynamic is a very different matter. Its complexity restricts our ability to comprehend it. Its dynamism means that its sands are always shifting. It is likely to come up with nasty surprises which we failed to foresee.

Do you see a parallel here with Perrow's Normal Accident Theory (Chapter 3)? Perrow was addressing the probability of the failure of systems of operation. He found that probability of failure was a function of two factors, of the comprehensibility and degree of insulation (our terms) in the system. The probability was lower in comprehensible and well-insulated systems. It was higher in incomprehensible and badly insulated systems.

- An uncontrolled situation which is simple and static is comprehensible
- An uncontrolled situation which is complex and dynamic is less comprehensible, perhaps almost incomprehensible

- A decision will not be vulnerable to small changes in an uncontrolled situation from which it is well-insulated
- A decision will be vulnerable to small changes in an uncontrolled situation from which it is not well-insulated. If it is closely coupled to those changes, they will have an immediate knock-on effect

What does this all mean? It means that Perrow's Normal Accident theory, a theory of system failure, applies to the system of decision-taking. As you might expect.

It also means that, however well we control the controlled situation, our decision remains, to a greater or a lesser degree, at the mercy of the uncontrolled factors. We know that the best laid schemes of mice and men may gang awry.[81] And now we can see why.

What do we do? We do our best.

We know that the Gods will have their fun if they want it. But now that we can see what they are up to, we can give them a run for their money.

RULES OF COMBAT WITH UNCONTROLLED FACTORS

1. We make maximum effort to identify all the important factors in the uncontrolled situation
2. We assess our insulation from each uncontrolled factor
3. We assess the comprehensibility of each uncontrolled factor
4. We make our best endeavours to improve our insulation from each uncontrolled factor
5. We make our best endeavours to improve our comprehension of each uncontrolled factor

81 'Gang awry' is fairly incomprehensible to English-speakers. The poet actually wrote 'gang aft a-gley', which is totally incomprehensible. Hence the customary misquotation.

6 We make contingency plans
7 We watch the uncontrolled situation like a hawk, ready to swoop and intercept any movement which may upset the course of our decision

MEANWHILE, BACK AT THE RANCH

Meanwhile our concern with the uncontrolled situation must not lead us to overlook the controlled situation. We have defined it as being totally under our control. If we are not too picky, this definition is not unreasonable; we are assuming that we know our business, that we are good managers, and that all in the company are committed to its well-being.

We will be sure, will we not, to avoid the obvious self-inflicted injury – complacency? We will not take the controlled situation for granted. In our tile factory, retribution will be swift if the data supplied by production or marketing is wrong, or if we fail to produce quality tiles of a superior design, or if we blunder elsewhere in the controlled situation. We must not neglect our controlled factors. *Complacency? Thunderbolts, Go!*

MENTAL FACTORS

In our tile factory the important factors were in the physical world. In many decisions the important factors are predominantly matters of perception; they are in the mental world. Should you accept this job offer, marry this person, have children, move house, get divorced?

Let us take an example of a decision, the factors of which are largely mental, and which also illustrates the difficulties of anticipating future feelings. We will take something nice and simple. Marriage. Should you? Or should you not?[82]

82 If in doubt, don't.

CASE STUDY 10.2
FACTORS OF THE WORLD OF MARRIAGE

At the Time of the Decision

The Imperative (we will assume) is to obtain long-term happiness.

The Premise is that, by marrying Mr/Ms Right, living together, having children, sharing life as a family, will yield the good feelings which will achieve the imperative.

The Decision is to marry Mr/Ms Right if he/she comes along.

The Original Criteria which distinguish Mr/Ms Right from Mr/Ms Notright are, (e.g.)[83]:

Factor 1 Personality
Factor 2 Beddability
Factor 3 Common-sense
Factor 4 Financial sense
Factor 5 Pleasant in-laws.

Mr/Ms Right meets these original criteria. Ding Dong. Wedding bells.

In the Future

New factors have probably appeared (eg):

New Criteria

Factor 6 Common Interests

New Factors

Factor 7 Tempting job offer from overseas
Factor 8 Childcare conflict
Factor 9 The in-laws wish to move in
Factor 10 New hobby
Factor 11 New dolly/hunk.

83 Simplification of our earlier list (Chapter 8).

These six new factors, these important factors, lie more in the mental than the physical world. They are largely in the uncontrolled situation, not under our control. They were not present, nor easily foreseeable, at the time of the original decision. But they are present now, when the success of the decision is being evaluated. Worse, some or all of the original factors may no longer hold good.

We reach the same conclusion as in the case of tile manufacture. The uncontrolled situation has enormous power. It can make or break our decision. The factors of the world, whether mental or physical, have us by the vitals.

CONCLUSIONS: FACTORS IN THE EXTERNAL SITUATION

We have now reviewed the factors in the external situation, that is in the world around us.

We have identified some of the steps in good decision-making:

1. Identify the factors important to the decision
2. Some of these factors will be under our control (controlled factors), and some will not (uncontrolled factors)
3. Remember that the uncontrolled factors may be very powerful. They may decide the outcome of our decision
4. Therefore, as ever, the Gods Rule

The external situation contains our problem. We have identified those of its factors which are important to our decision. To complete our study of the decision-making process, we must identify the important factors in the operation of our mind (Chapter 11), and we must review the available decision-making methods with equal care (Chapter 12).

If we understand decision-making, we can ensure that our process is sound. Then we may achieve a successful decision – if the Gods are kind. But if we don't understand the process, we will be in breach of the first rule of battle – know the rules of battle (Chapter 2). Then the Gods will have their fun, and it will be at our expense.

CHAPTER 11

DECISION-MAKING: THE MIND

FACTORS OF THE MIND

The mind is the satnav which must take us over the terrain to our destination. It must navigate us across the world, the external situation, to our decision. Unfortunately it is flawed. Some features of the terrain are missing from its database, and we are not sure which or how many. Some obstructions are too much for its hardware. And it is not particularly user-friendly, nor does it always do what it is told. So we have to keep an eye on it, if possible with a decent scale map to hand. Otherwise there is every chance that we will lose our way and end up in a farmyard or cul-de-sac or heading for disaster. All of which most of us have done.[84]

The better that we can understand how our satnav, our mind, works, the better that we can make good use of it. It plays two kinds of little tricks on us. The first is simply to fail to cope with the problem before it. The second is to try to deceive us. These little tricks can easily sabotage good decision-making. We must identify the bad factors of the mind in order to keep the little bastards out of the process. We must identify its good factors, those which will be helpful to our task, and marshal them into service. We have a continuous tussle with our satnav. Given half a chance, it will take charge of us. Therefore we must keep one eye on it and one on the map.

84 And probably not most, but all, who own a certain German car. Begins with a B.

THE MIND IN A NUTSHELL

A male pilot of a Boeing 747 has as much chance of understanding the details of his aircraft's workings as he has of understanding women. Fortunately he does not need to. He needs to know how to operate it. And he needs to know how to tell when bits of it are going wrong, and what to do when that happens. And that's it. Just like marriage. Only a lot easier.

It is the same with the mind. We need to know how to use it. And we need to know how, given half a chance, it will use us. And that's it.

We saw in Chapter 9 that good feelings were the deciding factor in making a decision. We saw in Chapter 10 how, with a little thought, we could base our expected good feelings on the real world. This is a sound foundation for a successful outcome to our decision. The problem with the bad factors of the mind is that they are capable of giving us the same good feelings, but based on flawed or wishful thinking.

When wishful thinking comes up against the real world, the real world wins every time. Those with an unlimited capacity for self-delusion will not perceive it thus. They will look in their mental mirror and see themselves successful and beautiful. May they be happy in their perception, and in their own little world. They will not look so good, nor perform so well, in the real world, and it is in the real world that accidents happen. Our battle with the Gods is real-life combat. Our everything is at stake. We cannot afford false perceptions. Wishful thinking is one thing, wishful flying is another.

We must never fool ourselves. Don't even think about it. Because for sure we can never fool the Gods.

THE BRAIN

Herodotus tells a story of the ancient Persians. They never, he said, took a decision sober but that they reviewed it drunk. Nor did they take a decision drunk but that they reviewed it sober. The ancient Persians sound a whole lot more fun than the present lot.

What were they up to? A sober decision is conscious decision, taken in the thinking corners of the brain. A drunken decision is a semi-conscious decision, less inhibited and therefore perhaps more in touch with our inner emotions. It is taken in other corners of the brain, the thoughtful corners being out to lunch.[85] Alcohol is technically a depressant, because it depresses the activity of the hypothalamus, an inhibiting corner where the learned conscience may live. So if too much alcohol is bad for you, so is too much thinking; it can drive you to drink.

The brain has two sides, left and right. The left side is logical and verbal, loosely words and details. The right side is creative and visual, loosely pictures and ideas. They are connected, usually better in women than in men.[86] One side or the other is usually dominant, and this is the one that the individual normally brings to the decision-making table. That is nearly all we need to know about the two sides of the brain.

Continuing in general terms, the brain has two methods of thinking, cognition and intuition. Cognition is the processing of data at the conscious level. Intuition is an unconscious process. Only the results of the process are posted up to the conscious level. The details of the process itself remain in the unconscious. Cognition and intuition each has an important role in the decision-making process. We will look at them both.

Note that intuition and cognition are not neatly confined to one side or the other, they take place all over the brain, as does sober thinking. Drunken thinking, as you would expect, staggers about unsystematically.

Before we discuss cognition and intuition, we must be clear about the brain's default mode. Whenever it is not in repose, whether blissful or plain indolent, or fully engaged in commanding action, it is trying to work out what is going on. It is trying to make sense of the world.

85 Hence *in vino veritas* (in wine the truth comes out).

86 Which may be why men can only do one thing at a time but ladies can do two or more. Exactly why they do them is a more delicate issue.

SENSE-MAKING

Sense-making is a happy expression because it immediately conveys its meaning. It provides a simple accurate picture. And the mind likes images and pictures.[87] One picture, goes the proverb, is worth a thousand words. And a thousand words may well be a serious underestimate of a cognitive psychologist's explanation of understanding.

The mind assumes order in the world and goes out to look for it. The fact that this is a pretty big assumption is neither here nor there, that is what the mind does. Actually it is difficult to see what else it could do, short of giving up and awaiting its fate.

The mind looks for connections, for some form of relationship between whatever it is thinking about at the time. Connections may be provided by pictures, patterns, narratives or frameworks. Frameworks may be provided by any system of thinking which puts the mental material in a context. Thus history offers a historical context, logic a logical context, maths a mathematical context, books on risk a risk context. Etc., lots and lots of etc's. There are as many contexts as human ingenuity can conceive, and as many ways of looking at the mental material. Clever minds have a large database of contexts and are good at selecting the most appropriate. We all know people who cannot go beyond the WIIFM context.[88] More about contexts in Chapter 13.

When the mind finds not connections, but contradictions, it becomes unhappy, sometimes very unhappy. 'Cognitive dissonance' is the name given to such contradictions. It is important and we will come to it in a minute.

The world is a biggish place and complicated; the mind is small, with lots to do and not much time to do it, so the poor thing is on a

87 An agent to whom I offered this book requested that I sum it up in one short sentence. He told me that the classic example was the film *Chicken Run*. He said that it was sold to Steven Spielburg with a single soundbite, 'The Great Escape. With Chickens.'

88 What's In It For Me.

loser.[89] Sensibly it tries to use shortcuts which have worked before (or so it believes). One form of shortcut is called heuristics. Clever people have heuristics which work and they are always learning new ones. Dopey people have heuristics which do not work, and the accident-prone self-evidently do not learn.

Unfortunately, the mind can only test one solution at a time. Further, as soon as it finds a satisfactory – and not necessarily a good, or the best – solution, it stops searching. Shortcuts thus have an obvious danger; if the mind accepts one, it promptly ceases to consider other possibilities. Soundbites and maxims may well offer good guidance in specific contexts. They may be a shortcut to disaster in others. 'Too many cooks spoil the broth.' 'Many hands make light work.' But, in a given context, which applies? As ever, context is king. Soundbites are beloved of politicians and admen because they convey a simple, immediately comprehensible, picture.

- 'Because you are worth it.' Oh, yes? No. Because they want your money
- 'Where do you want to go today?' Think about it. You are sat on your fat backside in front of a soulless screen. You are not going to the sunshine, nor the fresh air, nor to the gym, nor to a nice companion. You are going absolutely sweet nowhere. Loser

We know that we must beware of Greeks bearing gifts. Their information-age equivalents are conmen bearing soundbites. Do not be soundbitten; you will have been conned.

THE MIND

The mind is not a blank sheet of stationery, nor is it stationary. It is not a cold, calculating machine. It is a highly populated and dynamic world. It is into this whirling dervish of emotions, feelings, beliefs,

89 'Never take on anyone bigger than yourself' was the advice which I gave to my children, when they were small and engaged me in physical combat. Now that they are both large, not to mention the black belt, I follow my own advice very carefully.

needs, ideas, and the rest, that data enters; it is from this whirling dervish that decisions must emerge.

We do not fully understand it, and very probably we never will. We cannot reasonably expect the mind to have the capacity to explain itself. So, as is the case throughout life, we will just have to manage with whatever understanding we can obtain. Let us have a look at what is now generally agreed about the mind.

CONSCIOUS AND SUBCONSCIOUS

The mind is rather like the London Underground, only a lot cleaner. The conscious is what emerges from the tunnels into our view in the leafy suburbs (if only), where we are standing. The subconscious is what goes on underground. We cannot see it and our map is largely unreadable. With a bit of clever kit, we can manage an occasional peep into the darkness, listen to sounds, note what lights up when a train passes. We can put things in and see what emerges and guess what happened in the tunnels. But, by and large the subconscious is a mystery to us. Never mind, look on the bright side. We know that it is there, we can see what emerges; we are aware that, although we can hardly discern the contents, they nonetheless have powerful effects on us. We can cope. What we must always keep in mind is that the boundary between the subconscious and the conscious is permeable. Traffic can pass in both directions. And our ability to control that traffic is limited.

At birth the unconscious probably contains only some limited instincts. As we grow, it grows. More instincts develop. Experiences at the conscious level impact upon it. Very important, it learns. It learns morality; it learns religion; it learns social *mores*; it learns the interests of its host, the individual which is its life-support system; it can learn anything which achieves importance at the conscious level. These items of learning lodge at different levels in the subconscious.

Those at the highest levels, perhaps physical and mental skills, can be relatively easily raised to the conscious level and reconsidered, perhaps unlearned, modified and relearned. Those at the deepest levels, beliefs, values and other attachments, are matters of faith rather than understanding. Reconsidering and relearning them is a very different matter.

THE SUBCONSCIOUS

I am using the term 'subconscious' to refer to all levels below consciousness. Correctly the lowest levels should be referred to as the 'unconscious'. However, as there is no easy distinction between sub- and un-conscious, and as it is not important to our present interest, we will offer an insincere apology to pedants, and pass lightly over the possible differences between the two. Let us look at the traffic between the conscious and the subconscious.

Consider how you learn a mechanical skill. Take bicycling as an example. If unaided you will start by falling off. More likely your kind father will buy you a bicycle with stabilisers. This will enable you to learn to pedal and to steer, and to keep off Father's flower-beds and to avoid collision with nasty hard walls, all with minimal chance of falling off. When you have learned all this to Father's satisfaction, he will remove the stabilisers. You will then be able to give your full attention to the single task of remaining upright on two wheels, pedalling etc. having been learned. Soon you will be able to stay on without thinking about it. At this stage the skill has relocated from your conscious to your subconscious. You can ride without conscious effort; your subconscious is controlling your actions. However, if things go wrong, perhaps a sudden wobble, you can rapidly haul the skills up from the subconscious to the conscious in order to sort out the situation. Relatively simple mechanical skills are much better left to the subconscious; it can control our actions with much greater speed and dexterity than can conscious thought. How do you know when a skill has passed from the conscious to the unconscious?

Student pilots have to learn how to land an aircraft safely if the engine fails. (We are talking about small aircraft, not Boeing 747–400s.) In the mid–1960s the Jet Provost was the aircraft in which RAF pilots learned to fly. They were taught that in the event of engine failure, they should find an airfield (a good start), and should so arrange their flight that their aircraft ended up at 2500 feet, flying parallel to the runway on which they wished to land, and in the opposite direction. Then as, or after they passed their proposed touchdown point, to turn in towards it. That was the tricky bit. Turn in too early and you sailed over the touchdown point at great height and landed in the next county. Turn in too late and you landed several fields short and had to walk the rest. The wind was all-important in calculating when to turn in. In zero wind the rule of thumb was to fly past the touchdown point for ten seconds and then turn. For every knot of headwind down the runway (which of course was tailwind to the aircraft travelling in the opposite direction), one second was subtracted from the basic ten seconds. It was easy and it worked. The problems were that the aircraft was not always at precisely the right height, or at the right distance from the runway, or the wind was not behaving as instructed by the met man. Then rigid adherence to the rule of thumb would not land you on the runway but in the Squadron Commander's office. Rapid modification to the rule was needed, based on judgement. In the course of a few flights students learned to make this judgement with consistent success. When I felt that a student had mastered the process, I applied a simple test. As he glided along parallel to the runway, I started chatting lightly about this and that, and I kept going until he was safely established on his final approach. What I was doing was deliberately occupying his conscious mind with idle talk about life and witty observations about senior officers, to establish whether his subconscious had taken over the task of judging the forced landing. Note that I was not engaging the serious attention of his conscious (therefore no discussion of girls); this would have interfered with his unconscious activity. After landing I asked how he had chosen his turn-in point. The answer 'I don't really know, it just looked right' proved that his decision had been subconscious.

Robs Lamplough, pilot of Spitfires, Mustangs and just about everything else tells a story of his early motor-racing days. During the practices for the Los Angeles Times Grand Prix he was staying in a hotel with other racing-drivers. Before one particularly exciting bend there was a series of marker-boards. He asked Jim Clark which marker-board he used as a braking-point. Jim said he did not use a marker-board; he waited until it just looked right. Robs responded that Stirling Moss had written that he studied the track carefully and used a marker-board or track feature as a cue. Jim said that it was better to keep the accelerator firmly planted, and wait for intuition to lift the foot. In other words, the subconscious takes better decisions than the conscious, but only when and if it has the capability. Jim won lots of races, he was a World Champion.

These examples of the traffic across the conscious–subconscious boundary and of the power of subconscious judgement relate to the learning of motor skills. They are equally applicable to our present interest, the taking of decisions. In this context, subconscious judgements are usually known as intuitions. We will come to them shortly. First let us look at conscious judgements.

COGNITION

Cognition is the mind's conscious decision-taking tool. It involves awareness. 'I have this decision to take. I have this data. I need to consider these scenarios. I need to obtain that information from elsewhere, etc.' Feelings and intuitions will rise from the subconscious in the course of the process. They then become factors in the conscious mix, to be considered with all the others. If we make enough effort, we can keep cognition largely under our conscious control.

FAILINGS IN COGNITION

Unfortunately there are failings in the cognitive process. Some are its resident weaknesses. Some are incoming from the subconscious, bad factors disrupting cognition. We will have a look at the most important.

The first resident weakness is that cognition soon runs out of steam when faced with *complexity*. To comprehend complicated situations cognition has to break them down into smaller parts. Then it must reintegrate them into the whole. And it must hope that its understanding of the behaviour of the parts equates to an understanding of the behaviour of the whole. Sometimes it does. And sometimes it does not. Dissecting a beautiful woman would not explain why she is beautiful. Taking her clothes off would. And that is the key to understanding complexity. Not dissection but revelation. Pattern recognition.

Cognition's second resident weakness is its difficulty in the face of *excess information.* It has time constraints; it cannot examine each and every byte of information and seek to relate it to the matter under consideration. So how does it know what matters and what does not? How does it know where to start? It does not.

Cognition's third resident weakness is its difficulty when faced by a *paucity of information*, when there are not enough bytes to make a connection.

These three resident weaknesses are important mental limitations. When cognition comes up against one of them, it cannot perform. Then these weaknesses become bad factors, disrupting the thinking process. Cognition's first port of call will be a heuristic, if it can find one which is relevant. If not, it is likely to turn to intuition. This we will come to in a moment.

The next weaknesses of cognition are not limitations but failings in the mind's *modus operandi* (way of working).

The first is *cognitive dissonance*, which we mentioned earlier. Some use the term to refer only to contradiction between belief and information. This is unduly restrictive. The term accurately describes contradictions of any type, and we will use it in this, more useful, general sense.

Cognitive dissonance is an unpleasant feeling. It is often sufficiently unpleasant to prompt immediate efforts to resolve the contradiction. The easiest resolution is to reject one or other of the parties to the conflict. When one of the parties is a belief, it is likely to be much more powerful and durable than the item of evidence which conflicts with it. This item of evidence is then rejected, either bluntly or by means of considerable intellectual contortion. The obvious example is religion. Ask those who believe in an all-loving God to explain war or cancer or other ghastliness, and watch their contortions. Extremists will not even engage in discussion. They cannot tolerate any questioning of their beliefs. What psychologists call cognitive dissonance, they call blasphemy.

Mental commitment is one step down from belief. If it also involves a publicly pronounced plan of action, it will be accompanied by substantial social pressures: pressure to prove oneself right to stakeholders, colleagues, the boss, plus perhaps the pressure of the knowledge that money and resources have been invested in the plan. Contradictory evidence may invalidate the commitment.

The failure to address such contradictory evidence with an open mind is a classic planning error. It may well be the most common cause of self-inflicted disaster. It is a particular danger in military and other authoritarian situations. Giving the boss the message that his plan is a disaster is much more likely to terminate the messenger's career than the plan. Examples abound: Gallipoli, Arnhem, Anthony the Abysmal's invasion of Iraq. Unfortunately, some eminent military commentators regard cognitive dissonance as psychobabble, and therefore do not address its malignity.

Cognitive dissonance has several close relatives.

Cognitive lock is the refusal to look outside the existing mindset.

Cognitive block is less a refusal than an inability to look outside the mindset. Seeing, we are told, is believing. Equally important, to believe is to see. In other words, unless you are prepared to believe that something exists, you will not be able to see it. The antidote to cognitive block is open-mindedness.

Cognitive bias is more important because it may be unconscious. Max Bazerman, a Professor at Harvard Business School, reported one study of cognitive bias. Auditors employed by a company were less likely to find fault in its accounts than were independent auditors. In that case cognitive bias involved *framing bias*, putting the matter into the context most comfortable to existing beliefs or objectives. As we have seen, context is king, because the uncontrolled factors, not under our control, have the power to destroy any plan of action. To misidentify the context is also to misidentify the uncontrolled factors.

Framing bias is often used deliberately by those seeking to manipulate an outcome. It biases the subsequent decision. The classic example is the Labour Party's extraction of funds from lethargic trade unionists by requiring them to take positive action to opt out of paying; their natural inaction leaves them paying. Framing bias is intended to keep the funds rolling in, and it does. A similar proposition has been advanced by those who press us to donate our organs after death. Framing bias can be used unscrupulously in many ways. The inquiry into New Labour's Price-for-a-Peerage, one of the many great disgraces of Anthony the Abysmal's tenure of office, was framed not to inquire into New Labour's Price-for-a-Peerage, but to inquire into the anodyne and totally different subject of political fund-raising.

Priming is a form of framing bias. Smiley faces, warm coffee and other positive, feel-good stimuli can encourage positive and optimistic responses, where negative, draining stimuli can prompt negative responses. Hence the smiley faces of salesmen, politicians, tricksters and others seeking to secure your trust. Priming has recently been given a new name: *Nudge*. Whether witting or unwitting, and whatever it may be called, it can cook your decision-making books. Watch out for it, and boot it out.

And finally, feelings and cognition.

Feelings are obviously not a consideration in the cognitive workings of the mind of man, uniquely rational being and pinnacle of the evolutionary tree. We can think coldly and dispassionately when we put our mind to it. Or can we? Unfortunately, absolutely not. Feelings, like beliefs, force their way into the conscious. Given half a chance, they will try to take over, evicting contradictory evidence and other opposition.

INTUITION

Intuition is the second of the mind's decision-making tools. It is variously described as a gut feeling, an instinct, an inspiration, or as a message from God.

Intuition is an unconscious process. Only its outcome is conveyed to the conscious. The outcome arrives as a feeling, as a suggested response to a situation.

The workings of the process remain in the unconscious, unrevealed. Its validity cannot therefore be directly assessed. Its outcome can be assessed, and it has been the subject of much study. The first finding is that the unconscious is much faster than the conscious. It can command a mental or physical response before the conscious knows what is happening. This has survival value in that the response to danger – a coiled snake or a hot pan or whatever – is almost instantaneous, leaving the conscious to catch up later. It may be disadvantageous to thoughtful decision-making. If the intuitive answer arrives in the conscious before the cognitive process is complete, it may influ ence the outcome of cognition. Of this potential bad factor the thinker must be aware.

We have already seen that cognition runs into difficulty if it has too much or too little information. Cognition does not have the processing power to manage a heavy information overload. Nor does it have within itself the means to bridge any large gap between inadequate information and the information needed for a soundly-based decision. If it is irresponsible, it makes a blind leap across such a gap with a wild guess. If it is responsible, it listens carefully to intuition, which may well already be knocking on the door with advice.

This dimension is important to our study of decision-making. Information may be anywhere from inadequate to excessive, with just right in the middle. At each extreme we must rely upon intuition. In the middle, cognition rules.

Inadequate Information	**Just Right Information**	**Excess Information**
Intuition Needed	Cognition Rules	Intuition Needed

There is a nice wrinkle to this. The excellent Gerd Gigerenzer, whom we met in Chapter 3, shows in his new book *Gut Feelings* that too much information is bad, not only for cognitive decisions, but also for intuitive decisions. This puts the boot firmly into the received wisdom that the more information the better. It also gives some support to my own theory that most academics are poor decision-makers because, being bright, they can remember lots and lots of stuff and can see every possible angle and every possible built-in-but. They disappear into deep catacombs of complexity, never to emerge with a decision.

Gigerenzer and others convincingly show that intuition is often a better decision-maker than cognition. Unfortunately, complex real-life situations – get married – obtain promotion – change jobs – move house – emigrate – involve, not a choice between cognition and intuition, but the use of both. We are back to the tussle between the heart and the head. And we are back to context. In the matter of marriage, we would be ill-advised to allow the head to dictate our choice. In the matter of obtaining promotion we would be ill-advised to pass total control to the heart. Your heart may wish to inform your boss that he/she is a dictatorial worm/ess. Your head would counsel against mentioning it.

In contexts in which we have long experience, we are well-advised to listen most carefully to intuition, because it soaks up experience; it learns. A pilot with umpteen years of experience picks up tiny cues which totally escape beginners. This enables him not only better to see what is going on, but also to foresee what is likely to happen. Hence the definition of an ace: a pilot who uses his superior skills to avoid using his superior skills.

So should we always listen to the voice of experience? No. Experience, as you would expect, brings with it a built-in-but. A leader's record of success may well be based on his/her possession of better intuition, better speculation of the future, than his/her colleagues. This will give him/her a strong belief in his/her judgement. And strong beliefs tend to deal with cognitive dissonance by kicking out conflicting evidence. This is the classic over-reaching of a leader who is set for a fall.[90] Consider how many great leaders have crashed and burned. For example, Margaret Thatcher tested the idea of the poll tax

90 Hubris. O dear.

against reality in Scotland. It was a disaster, but she would not accept this conflicting evidence. She pressed on with similar measures in England. The rest is history.

There is a simple answer. Honesty. Cognition must always be honest with itself.

It must examine the incoming factor. Is it hard data, or a genuine intuition, based on experience and some evidence? If so, it is worthy of respectful consideration.

Is the incoming factor a belief seeking to prove itself right? Is it a guess, based on no experience and no evidence? Is it a feeling of piety or hope? These are all bad factors. Boot them out.

The subconscious is always seeking its own advantage, whatever makes it feel good. The conscious may listen, but it must be very careful to decide what is best for its owner.

'Sleep on it' is often good advice. This gives time for make-believe worlds to evaporate, for emotions to cool, for temporary highs or lows to subside, for the subconscious processes to refine their conclusions, and for the cognitive processes to chew over the situation more carefully.

There is no way of testing the workings of intuition; they are unknown. What can and must be done is to test the outcomes of intuition. The reality test of a decision is the success of the action which it has suggested.

A reality test is an honesty test. It behoves cognition to adhere unflinchingly to honesty in its acceptance and processing of data, in its acceptance and processing of messages from the unconscious, and in its own workings. Do not live in a make-believe world. The Gods operate in the real world. Their retribution will be in the real world. And sooner, not later.

CONCLUSIONS: FACTORS OF THE MIND

1 Our mind has limited capacity and limited ability
2 Our mind has strengths which may work for us
3 Our mind has weaknesses which may work against us

4 We need to keep an eye on all of them
5 Our mind is continuously trying to make sense of the world. It may get it wrong
6 The mind does not like contradictions. It will seek to resolve them. It may get it wrong
7 Cognition cannot cope with too much or too little information
8 Intuition is very fast
9 We cannot see the workings of intuition, only the products
10 Intuition provides brilliant insights and bad guesses and everything in-between.
11 The products of intuition must be tested very carefully against reality
12 The mind given half a chance, will lead you up the garden-path. Don't go

CHAPTER 12

DECISION-MAKING: THE METHODS

THE THREE METHODS

The prime requirement for a happy and successful life is the avoidance of extreme adverse events. Our lives are informed by our decision-making. For that reason we have been looking carefully at the decision-making process. We are using the analogy of a journey: the destination is our objective; the terrain is the external situation – the world; the satnav is the internal situation – our mind; the vehicle/s is/are our chosen decision-making method/s. We will now complete our study with a brief look at the vehicles available.

We have a choice of vehicles, sturdy or speedy, simple or complex. They will take us all, or part, of the way across the terrain to our destination. With difficult decisions we will have very little idea of the terrain or of what we may meet on the way. Which vehicle should we take? Or should we ignore them, and go on foot? The answer is fairly obvious. From your knowledge of the terrain, make the best choice that you can. Set out, and if you get stuck, dump it and try something else.

We saw in Chapter 9 that decision-making is not an objective, scientific process. At its heart is a subjective judgement. Underlying our purchase of a Batmobile was the judgement that speed and showing-off was more important to us than comfort or economy. This necessity to choose between unlikes, that is between qualities, or between quantities which are measured in different units, is the fundamental reason for which the process of decision-making defies science.

There are umpteen decision-making methods, many of them proprietary. Some are helpful in certain situations. None constitutes a general solution to the problem of decision-making. The best of them bring a useful discipline to thinking, helping to ensure that all crucial factors and all relationships are considered. The methods broadly fall into three categories, pathways, scorecards and mathematical models.

Mathematical models are for mathematicians. So if you are not, don't.

Pathway methods are usually called trees, because they branch off in all directions.

Scorecards usually take the form of a matrix. We will take one as our first example; it gives us some useful insights into practical decision-making.

SCORECARDS: LOCATION, LOCATION

The simplest form of a scorecard method is a list of pros and cons. Whichever list is longer wins. The first level of sophistication is to assign each item a weighting, to ensure that the more important factors have an appropriately greater impact on the score. The second level of sophistication is a matrix. Let's have a look at a practical example.

We are contemplating a house purchase. We have five *options*, five houses to choose from, all are within budget. Our judgement is that we must base our purchase decision on six *key factors:*

- Cost
- Ambience of house
- Ambience of its locality
- Parking
- Travel to work
- Local schools

These six key factors are in different units. To make them comparable, we allocate them numerical scores. We rate them 1 to 5, 5 being best. In costs, 5 is cheap and cheerful, 1 is posh and pricey. Here we go.

CASE STUDY 12.1. WHERE SHALL WE LIVE?

Key Factor	*House A*	*House B*	*House C*	*House D*	*House E*
Cost	5	5	4	3	1
Ambience/ House	4	5	3	4	2
Ambience/ Locality	2	1	3	3	5
Parking	1	4	5	5	2
Travel To Work	5	5	1	1	2
Local Schools	1	2	4	4	5
Totals	*18*	*22*	*20*	*20*	*17*

Top-scorer is House B, with 22 points. Decision taken, we will buy House B. Easy.

But then, oops. There is a mugging in the next street. We now notice that House B has the lowest score in *ambience of locality*. We cannot live there. We decide to forget House B and to buy either House C or D, both in second place with 20 points.

How do we choose between C and D? They have the same total points; this does not distinguish between them. We look closely and see they score the same on the last four measures, but House C scores 4 on Cost and 3 on A*mbience of the house*, whereas these scores are reversed with House D. On reflection we decide that 4 points of *Ambience* are more important to us than 4 points of *cost*. We decide to buy House D.

But then Red Ken digs up the underground to prepare for his Nuremberg Rally moment, the London Olympics. Houses C and D already have bad scores on *travel to work*; both involve the same underground line. And now further misery on the tube? Not b. likely. So we decide to buy the house with the next highest score. It is House A. Job done.

But then we notice its low score on *local schools.* Would we really entrust our Little Ermyntrude, the apple of our eye, to the dysfunctional functionaries of Trotsky Comp? Never. Never. Never.

This leaves only House E.

Thus we have decided to buy the lowest scoring house.

So much for decision by highest score. And so much for the scorecard method. Drat and double drat. Back to the drawing-board, Hoskins.

Later, ensconced in House E, enjoying the *ambience of the locality* (5/5) and thoughtified by a warming glass of red, we revisit our scorecard to see what went wrong.

We realise that we had underestimated the importance of *ambience of locality.* We should have deleted any house scoring less than 2. House B would have been out.

We realise that 4 points of *ambience of house* were more important to us than 4 points of *cost.* We should have downgraded the importance of *cost* by halving the points allocated to it.

We realise that we had not considered the impact on our daily lives of our underground route to work. Damn. We should have made it a factor on our scorecard, thereby adding an additional row. A low score would have eliminated Houses C and D.

We realise that any house with a rock-bottom score of 1 on any factor should be eliminated.

And then we realise that our priority in life is Little Ermyntrude. We should have made *local schools* the top row and the deciding factor. This would have immediately yielded the decision to buy House E. If one or more Houses had had the same score on *local schools* as House E, then we would have used their scorecard totals to choose between them.

We bought House E, the house with the lowest score. Does this lead us to conclude that the scorecard approach to decision-making is NBG?

No. The scorecard served its purpose well. It guided us to what we now consider to be the best decision. That is exactly the true purpose of a methodology of decision-making. It does not take the decision for us. It guides us through complexity and informs our judgement. It enables us to take the 'best' decision.

As we have already seen, decisions are judgements. Judgements are made by people. Not by methods, not by systems, not by computers. Wrong, you say, and offer examples of decisions taken by methods, etc. Not wrong, the author replies politely. Underlying all such systemised 'decisions' are two judgements made by people. The judgement of the bloke/blokess who decided the workings of the method, and your own judgement in deciding to allow the method to take the decision for you. Before you pass authority for decision-making to any system, ponder GIGO. Garbage In, Garbage Out. Ponder well.

PATHWAYS

In his book *Gut Feelings*, Gerd Gigerenzer gives a particularly useful example of a pathway method, and of the ponderous mathematical model which it replaced. He cites a study conducted at a hospital by researchers from the University of Michigan. They investigated the success of doctors at distinguishing a very serious heart condition from less serious heart conditions. They were not very good at it. The researchers devised a complex mathematical model involving fifty probabilities, complete with a special calculator. Results improved, but the doctors did not understand it, nor did they like it. The researchers withdrew it. They substituted a very simple pathway model of a type which Gigerenzer calls a fast and frugal tree. You will see why.

In his example (below) a patient has a suspected heart problem. The doctors have to decide quickly whether to put the patient into intensive or normal care.

Their first question relates to the most important indicator of a very serious condition, a change in the ST segment of the electrocardiogram. If the answer is yes, there is a change, that is the only diagnostic step needed; the patient is immediately admitted to intensive care. If the answer is no, diagnosis continues.

As you can see, a final decision may be reached at any stage, with no further factors considered. Hence fast and frugal.

It is obvious, but still worth reminding ourselves, that this method can be used only when the validity of the factors as decision-takers has been positively established.

CASE STUDY 12.2 FAST AND FRUGAL TREE: CORONARY CARE

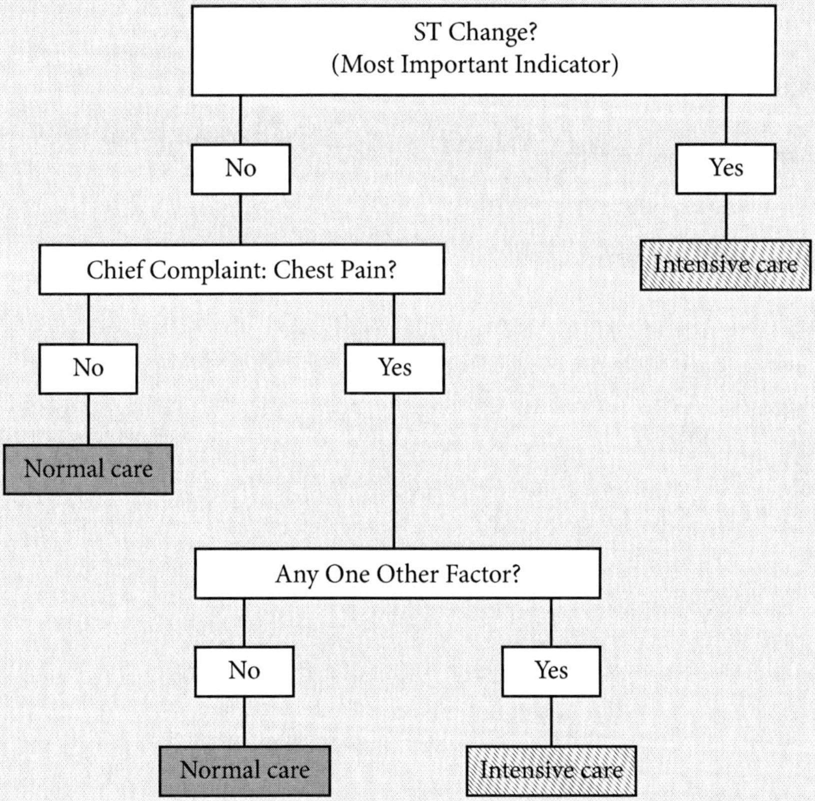

The fast and frugal tree out-performed the doctors' own judgements and the mathematical model, both in identifying the very serious cases and in reducing false alarms.

Note that to send non-serious cases to intensive care is not fail-safe. Intensive care exposes patients to a greater danger of hospital-borne diseases, from which they may die. As always, measures to reduce one risk introduce another. Built-in-but. Further, the unnecessary use of intensive care wastes resources, and ultimately deprives a serious case of the facilities.

Gigerenzer completes this revealing study by comparing the simplicity of his fast and frugal tree with the galloping complexity of a full decision-tree.

The number of exits or answers on a fast and frugal tree is the number of factors, plus one. Our coronary care example had three factors. It therefore has four exits.

A full decision tree examines a series of factors; each has two possible answers, usually yes or no. The number of exits or answers is two to the power of the number of factors; this in our example is two to the power of three, eight exits.

Taking an example with four factors, the fast and frugal tree would have four plus one, five exits/answers; the full decision-tree would have sixteen (two to the power of four). With 20 factors, the fast and frugal tree would have 21 exits; the decision tree 1,048,576. Decision trees are of least use where they are most wanted, in complex situations. The required quantity of input data gallops geometrically into stratospheric quantities. Even if the crunching is done in a computer, the problem of the accurate estimation of the input data rapidly escalates. It soon invalidates the process.

MATHEMATICAL MODELS

Mathematical models depend upon three big ifs.

- If all the variables can be correctly identified
- If the relationship between all of them can be correctly identified, and

- If that relationship can be accurately expressed in mathematical terms

But, before these three big ifs, come three even bigger problems.

- The first is the precondition that must be met. A mathematical model must be the decision-making method most appropriate to the situation
- The second is how you establish that this precondition is met.
- The third is how you know that the model which you constructed today is valid tomorrow. The answer is that you do not

The mathematics is mind-bending in its complexity. The modeller trying to sell his method may understand it, but the chances of the average bloke/ess understanding it is zero. And the chances of the average below-average bank director understanding it is even lower. Unfortunately mathematical models are a standard working tool in the world of finance. Time and again a wizard emerges to announce that he has modelled some aspect of it. Often they are initially successful. Then the sands shift and they sink. The classic case was the massive hedge fund, Long Term Capital Management. Founded in 1994 by supposedly the world's brightest and the best, including two Nobel Prize winners, it was spectacularly bust in 1998. Fearful that its failure would trigger financial meltdown, the US Treasury bailed it (and its management) out. Ring any bells? Northern Rock? US Loans and Savings? Lehman Bros, AIG, HBOS, etc. Most supremos of the industry do not understand complex models. All that they understand is that the models are bringing them vast bonuses. Therefore they do not enquire too deeply. So now, not only do we have a model which sooner or later will end in horrendous losses, we also have unsupervised traders. We have a situation the outcome of which we can forecast. Don't be there when it happens.

The supremos' disinterest is a failure of the third essential of decision-making, integrity. They are driven by the wrong imperative: whether wishful thinking or personal greed you may decide for yourself. But do remember that they usually give themselves bonuses for their handling of the crisis which they had manufactured. Remember that it is our money that they lose. Vote Malik, and we will have some plump faces on pikes.

A mathematical model was used in the Britain's foot-and-mouth disease crisis in 2001. Millions of animals were slaughtered. The modellers claimed to have solved the problem. Their many detractors say that the mathematical model of slaughter was not the appropriate decision-making method. They say that the correct solution was vaccination, not slaughter. This leaves unanswered the question of whether the model was itself correct. There you have a classic decision-making problem. How do you judge something which you do not understand, particularly if it has not been previously tested?

We are continually warned that 'past performance is not a guide to future performance'. Absolute balderdash. In the plural. Of course it is. Bad cooks consistently produce bad meals. Bad drivers consistently bust cars. Bad solicitors consistently drop you in the mire. Crooks consistently try to diddle you. Bad governments consistently bring disaster. No wonder Anthony the Abysmal was for ever chanting his mantra, 'People do not want to look back, they want to look forward'. He did not want them to look back over his past performance, because they would have known exactly what they had to look forward to: more disasters.

There are many decision-making mechanisms, the workings of which we do not understand. Never mind mathematical models, what about intuition? We use it all the time. There is only one real test of the validity of a mechanism: has it worked well in the past? If no, be jolly careful about using it now. If yes, the question becomes: is the present decision-making situation broadly similar to those in which it has previously worked well?

The warning 'past performance is not a guide to future performance' is plain wrong. Past performance is not a *guarantee* of future performance. It most certainly is a *guide,* and the best guide available. The warning is designed solely to transfer risk from the establishment, where it belongs, on to you, the investor. Its purpose is to cover that part of the establishment's anatomy on which kicks should be landed. Vote Malik and your every investment will come with a free heavy boot, or a pike. Your choice.

If you do not understand mathematical models, they should not be in your decision-making armoury. For you, they are not a method of risk management, they are a source of risk.

The same is true of all models. Let us end with a revealing and entertaining anecdote which Gigerenzer recounts in *Gut Feelings*, significantly on the very first page of his book.

CASE STUDY 12.3. BEWARE OF PROFESSORS BARING MODELS

A professor from Columbia University was struggling over whether to accept an offer from a rival university or to stay. His colleague took him aside and said, 'Just maximize your expected utility – you always write about doing this.' Exasperated the professor responded, 'Come on, this is serious' .

He uses the anecdote to illustrate the general supremacy of intuition and simplicity over agonising and complexity. I use it to strip models.[91]

CONCLUSIONS: DECISION-MAKING METHODS

1 Decision-making methods do not take decisions for you, (that is, if you have got any sense)
2 Pathway and scorecard methods, when appropriate to the situation, can assist you to analyse the situation. They can enhance your understanding and inform your decision

91 of their pretensions.

3 Proprietary methods of decision-making tend to come with weighty names; Multiple Regression, Multi-Criteria Decision-Making, et al. At heart most of them are scorecards. They may be useful in specific situations
4 Mathematical methods are OK, a) if you are mathematician and can understand them, and b) when they are right. Unfortunately, you find out that they are wrong only when whatever they are supporting collapses and dumps you in whatever you are trying to avoid
5 If anyone seeks to persuade you to use a decision-making method or to make an investment which, after their explanation, you do not understand, invite them to go forth. Do not encourage them to multiply

THE DOLLY MIXTURE SELECTION BOX THEORY OF DECISION-MAKING

Let us not be daunted by our own advice in conclusion 5 immediately above, but end our study of decision-making bravely. We will take the Garbage Can Lucky Dip Theory a step further. We will propose a brand new theory, the Dolly Mixture Selection Box Theory.

We have already seen that there are three core inputs to the decision-making process. These are reflected in the three compartments of our Dolly Mixture Selection Box.

In the first compartment is the external input, the key factors which we identify in the external situation, in the *world* surrounding the decision.

In the second compartment is the internal input, the mechanisms of the *mind* which we choose to bring to bear upon our decision-making problem: cognition, intuition, framing, heuristics, etc. We are also acutely aware of the mind's tendency to play tricks on us. The mind always tries to make itself feel good. It will try to subvert our decision-making process by pushing its beliefs, values and intuitions. It will seek to eliminate cognitive dissonance by framing and other biases and by whizzing its favourite heuristics up from the subconscious into the cognitive process. So we have to watch it like an hawk. It must work for us, if we are not to be deceived into working for it.

In the third compartment are our *decision-making methods.*

The Dolly Mixture Selection Box Theory states that, when faced with a decision, we should thoughtfully review the contents of the selection box, and carefully select from each compartment those dollies most appropriate to the problem before us. Also, when making our selection from the second compartment, we should take great care to prevent the mind from infiltrating any of its nasty little tricks into our selection. These tricks are Bad Dollies. We must watch out for them, and keep the little bastards firmly locked out of the process.

Thus, unlike the lucky dip of the Garbage Can Theory, we take thoughtful charge of the process. We carefully extract from the selection box, those dollies which will be most helpful to the problem before us. These are the Good Dollies. And we will most definitely not let any Bad Dollies come to our party and spoil it. Good decisions begin with Good Dollies.

And there you have the Dolly Mixture Selection Box Theory (patent applied for).[92] Most sweets will make you fat, dumb and happy. The Dolly Mixture Selection Box will make you an ace decision-maker, slim, bright and happy. Be the envy of your friends. Be amongst the first to buy.

92 Only joking.

INTRODUCTION TO PART IV
SAFETY IN SIGHT

CHAPTERS 13–15

Part IV completes our study of the prevention of failure.

In Chapter 13 we discuss strategic thinking, the processes by which we achieve the best possible understanding of the total situation.

In Chapter 14 we see that our exposure to most risks is not under our own control, but under the control of hierarchies above us. We study their conduct. We do not like what we see. We observe that their chief concern is to control the risks to themselves.

In Chapter 15 we draw on our studies to propose a methodology, seven essential steps in the analysis of risk. We observe that the availability of good information is paramount to good decisions, and thus to good risk management. We note that freedom of information often conflicts with the agenda of the hierarchies.

We conclude that we can greatly enhance our lives by a more informed management of its risks. We will have a much better chance of avoiding unhappiness. Therefore we will have a much better chance of being happier. We will certainly have more fun. We may even live longer.

CHAPTER 13

THINKING – JOIN UP NOW

JOIN UP NOW

We will see in Chapter 14 that our hierarchies are little to be admired. They have one set of rules for us and another set for themselves. Our noses to the grindstone, their snouts in the trough. Bastards. Let us set them a good example. We will apply our dictats to ourselves. We will advocate joined-up thinking, and we will join up our own thinking.

We will look again at the situation which we seek to manage. We will take a strategic view of it. In other words we will join up our thinking around it. Why? Because if our analysis of the situation fails to consider important contexts, or omits crucial factors, our understanding of the overall picture will be incomplete. There will be holes in our thinking, inviting entrances for thunderbolts. If we close the holes, the Gods will have thunderbolts to spare. With luck they will put them to good use. See immediately above (hierarchies).

STRATEGY

Big word, strategy. What does it mean?

A review of its many definitions is neither entertaining nor edifying. Let us instead go directly to the best definition. It was coined by a great strategist, perhaps the biggest ever. Sun Zu? Clausewitz? De Saxe? Schelling? Michael Howard? Great strategists all. But no. The footnote gives the identity of the well-rounded strategist who said:

'When you have a problem, you sit on your fat ass and think it out. Then you get off your fat ass and sort it out'. [93]

Beat that for a definition of strategic methodology. Ms Parton identifies the two key elements, informed decision and action, and she undertakes them both herself. Hurrah. Not for her the lofty view that strategy is the preserve of the elite higher ranks (along with the credit when it is a success, and spin and lies when it is a disaster, and along with the knighthoods and medals and perhaps the Presidency of the European Union), and that the dirtying of hands, losing of limbs, and dying, is for the PBI.[94]

We will not enter into that hardy perennial, the generally pointless debate on the dividing line between strategy and tactics. However we will note the difference between the view from the ground and from the air. The battle is fought on the ground. This book offers you a reconnaissance flight, an overview of our defences and of the surrounding enemy forces. The higher we climb, the broader the picture which unfolds. We can see the wider terrain, the dispositions of the forces on both sides, the gaps in defences, the position of reserves, the supply lines, the incoming weather, etc. But the higher that we climb, the less the detail which we can see.[95] We can see the woods, but not the individual trees. To win the battle, we need to see both. We know that gaps in our thinking or in our actions will translate into gaps in our defences. This book seeks to join up our thinking. Only the foot-soldier can see the individual risks around him/her. That's you. Only you can join up your actions.

Strategy is the process of encirclement, leaving no gaps. There are two loops which we need to close. The first loop is that around the key task of decision-making. This we have already closed with our comprehensive Dolly Mixture Selection Box Methodology (Chapter 12). The second loop is around the problem situation in which we find ourselves. To close it we must achieve an understanding of the total

93 Dolly Parton, television programme on her life.
94 Poor Bloody Infantry. Or Poor Bloody Pilots for that matter.
95 Built-in-but.

situation. There are two steps. The first is to frame it in different contexts. The second is to identify all of its relevant factors, and to rank them in ascending order of importance: *relevant*, *important* and *killer*.

By the end of this chapter we will have joined up both essential loops in our thinking. We may be the PBI, but we will also be strategists.

First, let us look at what we cannot do. Then at what we can do, and how to do it.

BEYOND OUR POWERS

We cannot change the past. We cannot control the present. And we can neither control nor forecast the future. So, are we on a loser?

No.

We have just got off to a good start in our strategy of risk management. We have identified what we cannot do. We will therefore neither agonise about it, nor will we waste energy attempting the impossible. We will concentrate on the possible.

BEYOND OUR POWERS: CONTROL

If we apply ourselves, we can achieve a degree of control of factors within our immediate environment. We cannot achieve total control because, as we saw in our Revisionist Normal Accident Theory (Chapter 3), risks are inherent in life, in assets and in activity. The further the factors from our immediate environment, the less our control; it rapidly diminishes to zero.

The maintenance of control is a continuous, unending task, because the world is for ever changing. Sands are always shifting. Every element, person, group and organisation, is continuously seeking its own advantage, and is also changing in its character. Many elements will be predatory. Some will be actively hostile. Even the most powerful organisations in the world cannot impose their will on their environment. During the Cold War, the ruthless Politburo was unable to eliminate all dissidence, or to prevent, before it murdered them, a number of very heroic Russians from working for a better cause than totalitarianism. Whether the USA has been any more successful in its attempts at world control is a moot point. Its active support of evil regimes and oppressive countries has lost it the sympathies of many, perhaps most, of its natural allies, as well as earning it the enmity of the world's vast Muslim community. It cannot prevent its enemies hitting back as best they can; 9/11 was in response to US foreign policy; the assassination of John Kennedy is thought to have been in response to US attempts to assassinate Castro. In the CIA's terms, blowback. In our terms, a foreseeable consequence.

Nor was the USA able totally to control its thermonuclear armoury. It was not able to prevent accidents and near disasters. To put this in perspective, the USA is arguably the most accomplished manager of large, complex operations in the history of the world, albeit only on a good day. Scott Sagan's fascinating study, *The Limits of Safety,* catalogues a series of US nuclear mishaps, excursions to the very brink of a Third World War. Remember that in those cheery days both sides had nuclear weapons in the air, only one order, or one fumble, away from mutually assured destruction. Sagan reports one incident worthy of a famous Flying Circus. An intruder climbing the fence at a front-line US Air Force base triggered an alert. War alarms sounded. Nuclear-armed aircraft lined up for take-off. They were stopped only by a clear-thinking officer who drove his car at them (25 October 1962). The intruder turned out to be a bear, a nice touch which must have kept the Gods laughing for weeks. Let us hope that the officer got a gong and a huge pension. I bet he didn't. Sagan's study is certainly not the whole story. Nuclear attack plans were held wherever they were needed. There is no doubt that some of them ended up in the wrong hands. It still goes on. On 29 August 2007 a B–52 flew across the United States supposedly carrying dummy bombs, in fact carrying live nuclear warheads.

In Britain, even with military security awareness, laptops disappear from cars with monotonous regularity. Without such awareness, confidential databases are scattered about by government departments with abandon, and obviously without affecting the bonuses of their bosses. No surprise there. And those are the ones that you hear about. Trust HM's present government? There should be a SNAFU[96] section in the honours list, with a SAPFU[97] section exclusive to New Labour. No one can control their environment, but there are precious few who cannot do better than that.

BEYOND OUR POWERS: FORECAST THE FUTURE

If you wish to give the Gods a laugh, try planning ahead.

We can identify possibilities. We cannot forecast which of them will be realised. We cannot even forecast the weather, and the weather does not involve human behaviour.[98] Anything which does, and that is most things, is well beyond our ken. Despite millions of expenditure and some very sharp brains, the outcome of the various acts of the pantomime which constitutes the US Presidential elections cannot be forecast (until after the event, when large numbers of experts find that they were right). On the world stage, even the most massive and momentous of events – the fall of the Berlin Wall, the collapse of the USSR, Saddam's invasion of Kuwait – come as unforecast surprises.

Similarly with financial forecasts. Possibilities can be foreseen, but not the actual path of events, unless it is unavoidable. Interestingly, even then any small element of uncertainty makes it extremely difficult to obtain positive action. The 2008 mortgage collapse and its global repercussions are an example of failure of Olympic magnitude. If we apply our decision model, we find failures both of integrity and of intelligence. Many of those who drove the process must have been able to foresee its inevitable outcome, but were driven by the imperative of their own profits. Banks and investment houses were similarly driven to overlook our prime rule of risk management, understand what you are doing. The collapse was forecast by experts more than

96 Situation Normal, All You Know What.

97 Surpassing All Previous You Know Whats.

98 Following the famous 'there ain't going to be no hurricane' incident, a meteorologist was berated on the radio for his profession's failings. He pleaded for sympathy. We meteorologists, he said, believe our forecasts. We go out without an umbrella and get wet.

twelve months before it occurred.[99] One witness claims unequivocally that he forecast it to the boards of banks; they could neither understand, nor did they wish to know.[100] Their failure to understand was a failure of the second essential of decision-making, intelligence. Their disinterest was a failure of the third essential, integrity. Their decision-making was driven by the wrong imperative. Solution: plump faces on pikes, *pour encourager les autres.*

James Montier, psychologist and investment banker, may not be hugely popular with his investment banker colleagues. His paper, *The Seven Sins of Fund Management,* is a hilarious and thoughtful debunk of the supposed mystique of investment management. At the head of his list of sins he places forecasting. He points out that it is well-known that investment advisers are hopeless at forecasting. This is hardly surprising; human beings cannot forecast the unknown, and investment advisers, traders, hedge-fund operators, *et al.* are members of the human race, if only just. Revealing memoirs suggest that many of their riches come not from forecasting the market, but from manipulating it. Montier's thesis is not the improper revelation of a business secret. Comparisons between investment advisers' performance and stock market indices are routinely published in newspapers.[101] Montier's paper is constructive critique of the behavioural biases which flaw the investment decision-making process. He, like us, is searching to improve the process. Happily for us, his task, investment management, is more difficult than our task, risk management.

Investment is, in general terms, a win–lose situation (your win is somebody else's loss), and thus a conflict with adversaries. It is also centred on human behaviour, that most complex and unpredictable of environments, which was beyond even Isaac Newton. 'I can calculate the movement of the stars but not the madness of

99 Amongst others by an eminent German banker. To my certain knowledge, because he forecast it to me after a game of tennis.

100 Channel 4 TV, *Dispatches*, Jon Moulton, managing partner of the private equity firm, Alchemy, 18 Feb 2008

101 Investment advisers make a lot of money; much of it they cream off their investors. They are not stupid. Their investors on the other hand....

men', he said, supposedly after he had lost lots in the South Sea Bubble (1720).[102]

Risk management is not centred in human behaviour but in system behaviour. That makes it easier. Further, risk management is not win–lose; if you avoid the accident, that does not mean that someone else has it. Certainly, safety costs money, so in that sense the system operator loses. But do we care? He runs the system; he takes money from it; system safety is his responsibility. We can advance safety by making it a win–win, by ensuring that the operator shares the benefits of the avoidance of accidents. It will keep his head off a pike.

WITHIN OUR POWERS

If it is beyond our powers either to control or to forecast events, then we must apply ourselves diligently to the next best. We must make every effort to understand both the world and ourselves. We need to identify the parts of the situation over which we have a measure of control. We need to make contingency plans to cover those parts over which we have little control.

UNDERSTANDING THE TOTAL SITUATION

We are now set to close the loop on our understanding of risk management. We have a good idea of the contents of all three compartments of the Dolly Mixture Selection Box. The world is the situation through which we wish to make our way; we have had a look at it. Our mind makes our decisions; we have seen how it works. We have reviewed the decision-making methods which we can utilise. Now we will briefly revisit all three, the total situation in which we find ourselves.

One would think that the more that one knows of a situation, the better. One would be wrong. We saw in Chapter 11 that our cognitive powers cannot cope with too much information. They sink their teeth into a large sticky toffee pudding of data and, munch as they may,

102 What was predictable was that the management sold their shares before the Bubble burst. *Le plus ca change, le plus les memes snouts dans le trough.*

come to a sticky halt as the toffee sets. There are many respectable academic studies which show that there is an optimum level of information, not too little and not too much. This optimum varies from person to person. The intuitive and risk-taking need less than the careful and risk-averse. A search for data is potentially unending: there will always be more that can be unearthed; it is a recipe, or an excuse, for inaction. No good. We want a decision and we want action. For a good decision it is not impenetrable mounds of information that we need; it is an insight, an understanding of the situation.

There are two steps to understanding. The first is the careful and correct framing of the problem, setting it in one or more appropriate contexts. The second is the correct identification of the important factors. We start with contexts.

CONTEXTS

Contexts range from universal down to personal. Some examples may help.

- A universal context is one in which a universal phenomenon is framed: evolution, the life cycle of birth to death, the repetitive cycle of growth and decay, Newton's Third Law (any action has an equal and opposite reaction), etc.
- A general context is a human rather than a universal viewpoint: regulatory, environmental, economic, political, social, historical, international or similar
- A local context is a locally prevailing framework: employer, employees, markets, transport, *mores*, a family, a group, etc.
- Finally, a personal context is just that, the viewpoint of a stakeholder

As well as spreading along the dimension from universal to personal, contexts spread over the dimension of all known disciplines: biology, physics, social sciences, etc.

Blimey, you say, this is getting heavy. And you consider skipping this section. Please don't, all will become clear very quickly. And you will find it useful. Let us take an example.

CONTEXTUAL ANALYSIS OF THE FUTURE OF AIRGROWNUP

This time we are not a tile factory, we are a small airline, AirGrownUp. Since our foundation five years ago, we have been growing steadily. We are profitable. We operate all over Great Britain. We are cheaper and much more congenial than the railways. Unlike the railways, our car parks are not run by a Mafia. Unlike coaches, we do not get stuck on congested roads, nor do our passengers get stuck in tiny toilets. Should we lease a fifth aeroplane to put on our existing routes? Let us frame this question in a number of different contexts.

CASE STUDY 13.1 CONTEXTUAL ANALYSIS OF THE FUTURE OF AIRGROWNUP

Contexts

Regulatory: If we become a bigger airline, will we be hit with a more expensive regulatory framework? *Yes.*

Market Cycle: Will the market continue to grow and thus enable us to fill another aeroplane? Or is it reaching a plateau, or a downturn? *The national economic outlook is poor, a downturn is probable.*

Social: Will the extra noise of extra flights arouse social opposition? *No.*

Historical: Did earlier internal airlines succeed or fail when they expanded on our routes? *They failed.*

Reaction: If we succeed in growing the market, will this attract other, stronger competitors? *Possible, perhaps probable.*

Employees: Will we be able to find additional staff who will work to our high standards? *In places, only with difficulty.*
Financial: If our expansion is successful, by how much will our shares increase in value? *20–30%.*
Board Relationships: Is there a consensus for or against? *No.*
Board Relationships: Is a serious Board split likely? *No.*

Analysis
The Board believes that these contexts provide enough information for it to reach a decision.
– It particularly notes from the historical context, that previous airlines had failed in expansion.
– It observes from the context of market growth, that a downturn is possible.
– It observes from the context of reaction, that predators may arrive.
– It observes from the context of employees, that it will face difficulties.
The Board concludes that this is not an appropriate time to expand.
– It observes from the financial context, that expansion is a desirable objective.
The Board concludes that it is committed to expansion in the future.
– The Board decides to start to reduce the risks involved in future expansion.
– The Board decides to embark on a programme to strengthen and secure its position with the local communities and with its customers. As a first step it resolves to support local emergency air ambulance services.

Robert Hagstrom is a long-time associate of Warren Buffett. We may safely conclude that he knows his onions. The only obvious fault of his excellent book, *Investing*, is its title; the book is not about investing. It is an exposition of the importance and value of setting problems in different contexts, the better to understand them. *Investing* is about thinking.

Hagstrom advocates an armoury of different weapons for thinking. He calls it a latticework of mental models. Not models in the precise sense in which we have been using the term, but more in the sense of ideas. He urges us to read widely, to draw fundamental ideas – universal truths – from physics, biology, social sciences, philosophy, psychology and literature. We called these frameworks 'universal contexts'.

We took his advice in considering the future of AirGrownUp. We borrowed from biology the concept of a repeating cycle of growth and decay. We applied it to our market and concluded that a downturn was imminent. We decided that a downturn was not a propitious moment for us to embark on expansion. Fast and frugal. Game over. Thank you, biology.

Note that if we had had very deep pockets, we might have come to the opposite conclusion. Lease another aeroplane, put excess capacity on the route, and run at a loss – we can afford it. Our objective? To put an enlarged infrastructure in place, and to bed down our new operation without the pressures of heavy demand. Only the dopiest of predators would arrive and open up an operation where there is already over-capacity. Later, when the market turns back up, we will already have our feet under the table, cutlery poised to have any incoming predator for breakfast. Whichever our decision, it will have been well-informed by the context of biology and its concept of the repeating life cycle.

Keep an eye on the newspapers. You will regularly see a statesman (not many of them) or leading businessman complimented on his ability 'to bring a fresh viewpoint' or 'to look at things a different way'. This clever clogs is framing the problem in a different context. He has probably read Hagstrom.

What has this all got to do with us? Answer: We are trying to understand our own complicated situation. We are trying to unlock its complexity. We will have a better chance if we have an armoury of different contexts in which to frame it. One of them may give us the means of unlocking it, a killer insight.

RELEVANT, IMPORTANT, AND KILLER FACTORS

Most real world situations are complex, potentially impenetrable. A vast number of factors have some potential for some degree of involvement, however small. We have to prioritise them if we are to manage them.

From the vast number, we identify those of importance. In Chapter 10 we called them important or key factors. Now we will rank them in order of importance: *relevant*, *important* and *killer*. It is essential to differentiate *important* from *killer* factors. A *killer* factor is one which could make or break the success of our endeavours. As we have massively reduced the number of factors, we may hope that they will now be cognitively or intuitively manageable.

- *Relevant* factors we keep an eye on
- *Important* factors we keep a firm hand on
- *Killer* factors we grasp by the throat

Remember that we are now addressing, not just the external situation, but the total situation: the world, the mind, the decision-making methods, all three compartments of the Dolly Mixture Selection Box. Killer factors are more likely to be in the external world than in the mind or in a decision-making method. But the mind or a decision-making method may have a crucial weakness or a great strength.

In our house purchase example, we identified six key factors (Chapter 12). Then we realised that one of them, Little Ermyntrude's school, was overwhelmingly important, a killer factor.

Not all situations contain a killer factor. Some contain one. Some contain more. We must most earnestly strive to identify any that the situation does contain. Any one will give us success. Worse, any one, if overlooked, will almost certainly turn on us, and kill our chances of successful action. 'The Gods have no rage, nor Hell a fury, like a killer factor scorned,' said the Bard, (nearly). The overlooking of a killer factor is a self-inflicted injury.

A killer factor may be our great strength or our great weakness. Either way we must be cognisant of it and address it. Remember that we are in a battle with the Gods. They are all-seeing. They will have identified all the killer factors. Our great weaknesses are the easy ways through our defences. Equally, our great strengths are also our potential weaknesses. If our strong points fall, we fall. Of this the Gods are aware. And so too must we be.

Consider war. Both sides will be searching desperately for a decisive weapon, whilst seeking to deny such a weapon to the enemy.[103] They will both be searching for decisive weaknesses in the enemy's defences, whilst seeking to close decisive weaknesses in their own defences. Examples of killer factors in the Second World War? Obviously the atomic bomb. Also, perhaps, the British mainland chain of radar defences in the Battle of Britain, British possession of Enigma.[104] Killer weaknesses: unbelievable deficiencies in British equipment and training in 1939, hence Dunkirk – hurrah for our politicians and our General Staff[105] – (ring any present-day bells? body armour? roadside bombs and thin-skinned Land Rovers? RAF Hercules without fire-suppressant fuel tanks? RAF Nimrods with known fatal fault?); failure of Germans to fit the Me 109 with drop-tanks in the Battle of Britain; German invasion of Russia; Japanese attack on Pearl Harbour, bringing the US into the war. More recent examples: Mr Ratner's unfortunate description of his gift products (1991);[106] the American disbanding of the Iraqi army (2003–4), etc. Discuss forever. The point is clear. Killer factors can be decisive. They can win, or lose, the day.

103 Hilaire Belloc was on to this: 'Whatever happens we have got
The Maxim Gun, and they have not.'

104 German secret coding device.

105 The admirable Colonel Wintle's book has one of the best opening paragraphs ever. Paraphrasing: 'I have fought against clots all my life. No doubt I shall die of one.' (See endnotes).

106 'Crap'.

THE ENEMIES OF UNDERSTANDING

Enemies may be declared or undeclared. Declared enemies are to be preferred. This is not to say that they are more likeable. But they are more honourable. And, more important, they are easier to deal with.

Most enemies of understanding are undeclared. Those who set out to deceive are unlikely to declare their intentions. Most passionately deny that they are attempting to deceive you, but they would, wouldn't they? They come in two forms. Those who have successfully deceived themselves. And those who are knowingly deceitful. The former are stupid. The latter may not be.

Both are likely to proclaim that they are working for your best interests. They are not. They are pursuing their own imperatives; they are following their own agenda. The clever ones, and unfortunately some are very clever, obtain a position of power. This gives a massive gearing to their ability to pursue their agenda. Often it gives them the all-expenses champagne lifestyle which, for most of them, is the most important part of the agenda. Where do you find them? In positions of power. In far too many governments. Try to think of any government which you would trust. Start with black Africa. Russia? China? Work your way round the world, take in South America, the Middle East, the Arab World, Pakistan, India, the Mediterranean countries. Then come closer to home, to the USA and Europe. Contemporary Britain? The European Union? Which of these would you willingly trust with your freedom of speech? With your healthcare? With your savings? With your children's future? Worrying, is it not?

There is an argument that of these factors, the keystone – the killer factor – is freedom of speech. Its most important benefit is in the organisation of dissent and opposition, and thus in the removal of the establishment. That is not the issue which we are addressing. We are not advocating revolution. We are concerned with communication. Without the free flow of facts and information, the hidden agendas remain hidden. Prime Ministers are able to take countries to war. Bankers are able to take us to the cleaners. Mayors are able to take us for a ride.

Our joined-up thinking is entirely dependent upon a free flow of facts and information. Without it we cannot achieve understanding. We cannot take well-informed decisions. We cannot manage our own risks. We have to put our trust in those whom we know to be untrustworthy.

In the next chapter, we will identify the risks posed to us by undeclared enemies, by hierarchies, regulators, and well-intentioned idiots.

UNDERSTANDING THE TOTAL SITUATION: CONCLUSIONS

1 The total situation is the one which we are seeking to manage with success
2 The better we understand the total situation, the greater will be our success
3 Our first step is contextual analysis. We frame the total situation in different contexts, thereby to analyse it from different viewpoints
4 Our second step is to deconstruct the situation into its constituent parts
5 We examine these constituent parts, seeking to identify *related*, *important* and *killer* factors
6 Thus we seek to close the loop around the situation, and to make a well-founded analysis of its every aspect
7 We have (we hope) arrived at a strategic solution
8 We have probably done better than the politicians and General Staff
9 Make that very probably

CHAPTER 14

THE ENEMIES WITHIN

LEADING TO DISASTER

Our exposure to the gravest risks is not in our own hands, but in those of our leaders.

They are responsible for protecting us against human enemies, whether from within our own country or from elsewhere. They are responsible for protecting us from the risks posed by organisations upon which we must rely: airlines, motor manufacturers, railways, drug companies, food chains, auditors, etc. They control the nation's responsibilities, for our health service, for the care of our elderly, for the nurturing and education of our children. They are responsible for the integrity of our body politic, for the protection of our national values, and for the well-being and good reputation of our nation. They have responsibility never to involve the nation and its servicemen in war, save when the nation itself is seriously threatened. Over these grave sources of risk, we have virtually no control. We must rely upon our leaders.

For the detailed management of most of these risks we must rely upon the regulatory regime: laws, regulations and regulators. Their task is enormous. Regulators, even if they know where to begin, never know where to end. In the final analysis, regulation can never succeed. Regulation cannot guarantee responsible conduct. It can only punish irresponsible conduct.

Our leaders and our regulators have an impossible task. They cannot guarantee our security and our safety. They know this. They therefore turn their attention to their own well-being, and to guaranteeing their own safety from retribution. They work less for us and more for themselves. They become double agents.

They are the Enemies Within.

From our leaders and our regulators, may the Gods deliver us.

IN PRAISE OF TREASON

Treason, we were taught at school, is a Bad Thing. Treason, we were told, is about who rules the land.

Actually it is not. It is about the benefits of ruling the land, and who has those benefits. To be fair, in the good old days it was occasionally about who ruled. Charles I managed to upset enough people with his taxation policy that they cut his head off. We may hope that this most excellent precedent makes an early reappearance, backdated in keeping with the former Chancellor's enactments. By all accounts Charles' successor, Oliver Cromwell, was not motivated by the perks of the job. He remains something of an exception.

The modern equivalent of the historic monarch is the Prime Minister or president. Try to name more than one who in modern times has ruled his/her country with the sustained intention of bringing about beneficial change for its inhabitants. The key words are 'sustained' and 'inhabitants'. Many start with ideological intentions. Most are brought up short by the combination of practical difficulties and active opposition. Then watch the sands shift. Their energies are refocused on feathering their own nest, and on remaining in power. They enjoy its perks. Excellent living, adulation, near-absolute power, status, bodyguards, index-linked pension, motorcade, free hols with other nobs, armoured car and free shopping for the wife[107]. All at taxpayers' or someone else's expense. Then retirement into a cushy number and a book advance of several millions for an

107 This makes for an interesting risk calculation. There is a shortage of armoured cars. Assume one attack on a minister in twenty years, and ten per week on British servicemen, 10,000 in twenty years. This makes the Prime Minister's wife 10,000 times more valuable than a British soldier. Information which should be prominently displayed in every recruiting office.

imaginative work of autobiographical fiction extolling the benefits which they brought to the country.

If they bring so many benefits to the country, why are they so anal about maintaining their control of information? The answer is simple. The more information we have about them, the less we respect and believe them. And, more important, the less we trust them.

Trust is the glue which holds societies together. It also holds followers to their leaders. We are not over-concerned with what leaders say when they are seeking re-election; there surely *caveat suffragator* rules – let the voter beware. Nor are we concerned at what they may say in a negotiating or a conflict situation; there honesty equates to naivety. What we are concerned with is their decision-making. If we are to remain their followers we have a right to assure ourselves that they are capable of sound decision-making. We need to know the quality of the information upon which they are relying, the adequacy of their intelligence – that is their ability to process that information, and the imperatives to which they are working.

The most momentous decision which a Prime Minister can take is to send his countrymen to war. Notice that it is 'send', not 'lead'. Let us examine, in the context of our decision-making model, Britain's decision to go to war with Iraq in 2003.

BRITAIN'S DECISION TO GO TO WAR WITH IRAQ, 2003

The information on which the decision was allegedly based was Iraq's supposed possession of weapons of mass destruction (WMD) at 45 minutes' readiness. The evidence is that this information was strongly disputed by experts in our own intelligence staff, and flatly contradicted by Hans Blix, the UN Inspector. Not to mention the 1992 Iraqi defector who had reported the destruction of all Saddam's WMD.

There is considerable evidence that the decision was the product, not of a proper process of decision-making, but of an agreement between the US President and the British Prime Minister. This indicates the omission of the second vital step in decision-making, that of applying intelligence, i.e., of appropriately and adequately processing the information. *Post hoc* decision-justification is a classic mode of failure in decision-making.

The imperative was supposedly the need to prevent the deployment of these WMD against ourselves. When serious doubt was thrown upon the existence of WMD, a second imperative was fabricated – the need to eradicate terrorists in Iraq. I have studied terrorism for twenty years and I assure you that the suggestion that invading Iraq would reduce the terrorist threat to us was beyond ludicrous; it was certifiable stupidity. We can be very sure that our intelligence services are the opposite of stupid. Nonetheless our Secret Intelligence Service, MI6, supposedly supported this analysis, thereby guaranteeing that its pronouncements would not be trusted for a generation.

So here we now are, facing the most serious and intractable terrorist threat in our history – and one exacerbated by our own leaders, deeply mistrusting their judgement, observing their total lack of organisational skills, and doubting the reliability of our intelligence services. All of which originated in the process of decision-making by which Anthony the Abysmal sent our forces to war.

Why did the US really go to war after 9/11?

Framing the question in context of the bully of the block gives us the answer. While everybody is watching, some foreign Johnny sneaks up and kicks sand in his face. He does not see who it was. He needs to reassert his dominance. He is going to smack somebody, and hard. Who does nobody like? Who is hostile to his special friend? Who is sitting on invaluable assets? Who has he successfully smacked before?

And why did Britain go along with him? Because our Prime Minister chose to set the situation in a fantasy context, his own heroism. He saw Saddam as a Hitler, hell-bent on aggression, his opponents as ineffective Chamberlains, and himself as the great Churchill. Three crucial images. All wrong. Saddam was not an active threat to our land, nor had he formidably professional armed forces in superior numbers just across the channel. Chamberlain was not ineffective; he was actually very effective in buying us precious time.[108] Finally, Churchill was not a posturing poodle. He was a bulldog.

Later, God was cited as the authorising agent. There are all kinds of difficulties with this. First, the electorate would prefer its elected representatives to answer to them. Second, the electorate would prefer its leader to obtain his guidance from official sources and to heed the large, expensive and professional bureaucracy maintained for that precise purpose. Third, there are established mechanisms for the declaration of war which do not allow for their overriding by the deity. Whilst intuitions can indeed be informative, it is unimaginable that, on the matter of starting a war, one man's imagined direct line to God be substituted for the nation's due and diligent process of decision-making. Finally, there is no way of distinguishing God's message from a gut feeling. The decision to go to war and to kill tens of thousands of people should be taken very slowly, very carefully and very, very, reluctantly, that is, thoughtfully. In the brain, where the grey matter is located. Not in the gut, where something different is located.

The evidence suggests that the decision by the US President and his little coterie of clots to go to war was driven largely by their perceived imperative to respond to 9/11. They are said to have entrusted the formulation of the occupation plan to an individual who cut out the State Department and other experts.[109]

This echoes the style of Anthony the Abysmal who preferred government by sycophants on a sofa to government by proper processes in state offices.

What, you patiently ask, has all this to do with our subject? Has the author gone off at a tangent? The author hopes not. Our subject is the management of risk. The alleged tangent explores a very serious risk, one over which we have little control, and one which is supposedly in the hands of the established, experienced, expert mechanisms of

108 Even then we were not adequately prepared. There is a powerful argument that Churchill would have been better-advised to have bought time with an expedient degree of appeasement. Then we would have been better prepared for the inevitable.

109 That individual was described by General Tommy Franks as 'the f****** stupidest guy on the face of the earth'. Hurrah, say I, for General Tommy Franks and for plain speaking.

government. To what extent total control was usurped by one individual will doubtless be established by historians. That individual, who said that he felt the hand of history on his shoulder, will soon discover that it was on its way to his neck.

This case study shows that, at the heart of a catastrophically bad decision, lie failures of all three essentials of decision-making, of information, of intelligence and of imperative. It shows that the choice of the wrong contextual framework can bring failure. These failures are poison to good decision-making. Let us look for the antidotes.

DECIDE. ACT. GET BOTH RIGHT

Would that Mr Benjamin Franklin were still with us. America, which means all of us, desperately needs his intellect. Apart from that small detail, I would wish to ask him whether he would add self-inflicted injuries to his duo of certainties in this world (death and taxes). I am confident that he would agree that self-inflicted injury is the most certain route to disaster. I shall ask him when we meet.

A decision informs action. For success, we must get both right. In the matter of war, successful action means winning. Therefore we do not start wars which we cannot win. We do not support human rights to the extent that we attack Russia over Chechnya or China over Tibet. Similarly we do not occupy countries which we cannot successfully control, pacify, and make a better place for the inhabitants. And if we do, we effect a self-inflicted injury and a predictable disaster.

In our book, self-inflicted injury is the most common, most certain and most unforgivable ticket to disaster. We know how to avoid it. But how do we prevent those more ignorant, our elected leaders, from involving us in their self-inflicted disaster? Obviously we seek to avoid electing flawed leaders, but of that there is no guarantee. Our recourse, our antidote, must be the law. A lawyer-proof law, ensuring that there is no way by which the country can be committed to conflict without due process. Due process means that the essentials of the decision-making process are undertaken by the establishment and the bureaucracy, and that the process is audited by a committee of parliamentarians and of the judiciary. The Cold War pitted two massive bureaucracies against each other. The decision by either side to go to war had to pass through a great bureaucratic grinder. Doubtless each side had its hotheads but, given bureaucratic competence, they did not have the power to initiate a war. It may not have felt it, but when NATO and the Warsaw Pact were facing each other down, there was relatively little chance of intentional war. There was a greater chance of accidental war, but that was much reduced by the installation of a hotline (in August 1963).

Unfortunately, the same is not true of the threat from terrorist groups. They are not generally bureaucracies. Most are no more than groups of hotheads, under the direction of a loose cannon. Therefore their potential possession of powerful weaponry is to be taken very seriously. It is perhaps so serious that, for the first time ever, there will be true and effective international cooperation against terrorists. Up to now, terrorism has been seen as someone else's problem. Now it is emerging as everybody's problem. We can only hope.

Unfortunately, (again), when the leader of an otherwise estimable state, perhaps with a hotline to God, circumvents the bureaucracy, he circumvents much of the safety system. A cannon has broken loose on our own side.

Sleep well.

FREE THE FACTS

We have a Freedom of Information (FOI) Act. It is to be applauded. It works reasonably well, not least because it has teeth; transgressions can be punished. But, as you would expect, those affected have responded to reduce its impact. The sands have shifted. Important discussions have shifted out of offices and on to sofas. The written record has become less comprehensive. Occasionally records cannot be traced. Doubtless shredders have shredded.

Dependent on the officials and the departments involved, information may be freely imparted, or it may require committed and resolute extraction. The latter is more likely for sensitive matters, and the reality is that it will be too time-consuming for most enquirers. Which is what the recalcitrant officials intend.

Some years ago, I sought data from several governments on the extradition of terrorists. With good reason, all were reluctant to respond. The good reason is that extradition is an excellent mirror on the reality and effectiveness of anti-terrorist cooperation. The reality is that such cooperation was close to nil. Hence the reluctance. I ended up with data from only two governments, HMG and the Dutch. The Dutch because I was assisted by a brilliant, iconoclastic, boat rocking, civil servant. HMG only through the good offices of my MP, Richard Needham, who put down a parliamentary question. Not one government acceded to my FOI request. Most ignored it. The USA replied that US freedom of information did not extend to *that* information. Doubtless, given time I could have extracted the information from most. I did not have the time, and that was what they banked on.

As it stands, the FOI Act is a good start. Doubtless there will be a battle between those who would emasculate it and those who would give it sharper teeth. One improvement would be to disallow the establishment any defence for which it was unable to produce the written record. In the case of the accident to RAF Chinook ZD 576 discussed in Chapters 7 and 8, the RAF stated 'After considerable review (the reviewing officer) concluded from his own analysis ... .' The RAF was unable to produce either his considerable review or his resultant analysis. They were thus unable to produce evidence of their detail, their data, their quality, or indeed of their existence.

The Official Secrets Act is a powerful block on unwanted enquiries. Sometimes its use is fully justified. Mostly it is not. As in those many cases when the hierarchy equates its own secrets with those of the state. There the threat is not to the state, but to officials. Then they use the Official Secrets Act as an Official Egg Off our Face Act[110], responding to enquiries with a cry of Egg Off – could be worse. In their defence, perhaps at a certain level their incompetence should indeed be a state secret.

110 OEOF Act, perhaps with accompanying quango OfOeof.

There are many battlefields in the war for freedom of facts. The MoD does not think it right that coroners should comment when servicemen are killed by MoD or HMG organisational incompetence. Anthony the Abysmal's press secretary sought to control information Politburo-style, by bullying the media. He supplemented his bullying with bribes in the form of exclusives. His considerable success is greatly to the media's discredit; they failed to do their job. The press secretary was doing his, albeit by most unpleasant means. To the media's further discredit is their use of their powerful position to convey personal opinions in the guise of facts. The BBC has a clearly leftist bias in its general reporting of current and political affairs. Certainly it is not uncommon for items to sound like a PR release for a government department. 'The minister's action is being seen as a bold step which will lead to early measurable improvements.' Oh, no, it is not. It is not being seen as anything, because this is the first we have heard of it. The honest statement would be, 'The minister would like it to be seen as a bold step, and he is hoping like hell that whatever it leads to does not make him look an idiot.' The original sounded like a press release because that is what it was, read by a journalist lazy or complaisant or both. 'The BBC understands that blah blah … .' 'The BBC can reveal that the Prime Minister … .' Oh, no, it does not. But it sounds a lot better than 'The Duty Parrot is reading another government press release; this one says … .' It would be nice to think that journalists could be bound by a media equivalent of the Hippocratic Oath, sworn to convey facts and only facts. Fat chance. Journalism is story-telling and facts spoil stories.

Our only recourse is to pursue freedom for facts and to remember that facts do not lie, but politicians, hierarchies and journalists do.

HURRAH FOR WHISTLEBLOWERS

In theory, whistleblowers enjoy a measure of protection in many countries. In practice, they seriously endanger their employment, their prospects and their health. In Britain, they are told that 'Advice is available from ACAS'. Try it. The first problem is getting ACAS to answer the phone. 'All our operators are busy. Please try again later.' Keep trying. And then, hurrah, an answer. 'I wish to blow a whistle, please can you give me some advice?' 'Yes. You need to speak to a solicitor'. Gosh, what terrific advice, we would never have thought of that. So, before we have even started, the coils of the legal system are already encircling us. Our employer's system, probably both powerful and hostile, is to come. Your job will be in jeopardy. Fighting systems is not for the faint-hearted.[111]

A number of aircraft accidents were investigated in *Air Crash*, a series of excellent television documentaries. The cause of one airline accident was revealed by a whistleblower to be the failure to grease a vital control mechanism at the stipulated intervals. The controls of the aircraft failed, and it was lost with all on board. Discussing the case, Mary Schiavo, Former Inspector General US Department of Transportation, said that her advice to potential whistleblowers was: 'you need to know, you need to be prepared to find another line of work because you will not work in the industry, you will not work in the government.'

She added: 'In most cases it is almost impossible to be a whistleblower and survive in your career.'[112]

If that is the case in the US, land of freedom and of sharp-toothed lawyers, then in Britain the whistleblower must indeed be a brave person. HMG, far from being a staunch fighter for free speech, is arguably its most vengeful foe. There is a long list of government servants who have been either persecuted, prosecuted, or imprisoned for revealing information about illegal activities: rendition or torture, or about deranged activities: Bush advocating the bombing of the Al Jazeera broadcasting station, or simply about ministerial incompetence. In such cases the action is based on the Official Secrets Act, but the imperative which drives it is to conceal the illegal or brainless blunderings of an unworthy in the hierarchy. Dr David Kelly was

111 A strip cartoon set in the stone-age showed an unfortunate lying kicking and screaming. 'What's his problem?' 'It's his nervous system' 'He's done for. You can't fight a system.'

112 My attempts to verify this quote with the British agents for the series went unanswered. My thanks to Andrew Santoni of the US who responded to my son's web research request with this information.

hounded to death as a result of his whistleblowing over Iraq. If failure to take their opportunity to prevent his death were the criminal test of liability, Anthony the Abysmal and his press secretary would be facing homicide charges. They, of course, may leak freely without fear of retribution. Why? Easy. They make the rules. Two sets of rules, one for them, one for the rest of us.

A government whistleblower who, by revealing the ineptitude of the prime or other minister, makes him/her look stupid, will be pursued vindictively. An experienced intelligence investigator will be employed to identify the criminal purveyor of honest information. Unless he/she has been highly skilful, he/she will be found.

Why do such investigators pursue their quarry with relish?[113] Values. Senior civil servants of the old school are men of their word. Discuss with such men whistleblowers, or ambassadors – Sir Christopher Meyer, Craig Murray – whose honesty and plain-speaking has embarrassed the government, and to a man they set the case in the context of breach of trust. To them, the sanctity of a promise overrides any other question, including that of the public interest. Similarly, I asked the RAF's expert Chinook test pilot why he did not simply ignore the order not to give evidence to the investigator of the accident to RAF Chinook ZD 576. He had, he said, sworn an oath of allegiance and obedience. To him and to senior civil servants, their word is their imperative. For this they are to be admired, not criticised. But I think that they are wrong.

What is needed is a legal framework and mechanism to rule on such dilemmas. Given that HMG, other governments, the European Union, the World Bank, the UN, all the most powerful organisations on this earth have shown their willingness forcibly to protect their vested interests by gagging whistleblowers, such a mechanism will be a long time coming. Our MPs are coy about their expenses.[114] They obtained for themselves exemption from the Freedom of Information Act. They obtained the dismissal of the parliamentary watchdog, Elizabeth Filkins, for doing her job. Is the House of Commons a place of honesty?

113 Apart from the substantial fee.
114 I say coy. I think lying bastards.

Discuss. Pardon? Is it what? Try to be polite. Do we really expect MPs to provide genuine protection for whistleblowers? Do we really expect their masters actively to assist in the revelation of their own incompetence or misconduct? Our best hope may lie in web campaigns, where millions may be mobilised to support whistleblowers and to name and shame and pour scorn upon their persecutors.

A PLAGUE UPON US

It was supposedly Napoleon who called us a nation of shopkeepers. We are not. We are much worse than that. We are a nation of traffic wardens. Our favourite occupation is telling others what to do, and – with particular pleasure – what not to do. Add to that our traditional respect for the law, and there you have it. The ideal breeding ground for regulators. Britain, once Great.

There are three questions central to regulation.

1 Do we need it?
2 Where do we start?
3 Where do we stop?

Do we need it?

Could we not leave market forces to handle risks to consumers? Yes, with suitable incentives and sanctions, doubtless we could develop some kind of marketplace in which good practice fared better than bad practice. The theory nice in theory, rotten in practice. *All people and organisations will always try to get away with whatever they think they can get away.* (Malik's Second Law of Life).[115] The process would cost us many lives and much misery. Perhaps we can agree that there is no question: we must have regulation.

115 Malik's First Law of Life states: *All people and organisations will always pursue their own interests, as they see them at the time.* Malik's Laws listed in Appendix to Chapter 15.

Where do we start?

We might also agree that here also there is no question: regulation must start where the risks are highest. Risk, remember, is loss multiplied by probability of loss. Loss, we might say, is greatest when it is loss of life. We will start where many lives are at stake. Perhaps with air travel. Modern airliners carry 400 or more passengers. We will continue with other organisations that can kill, injure people, or destroy their health. Then we will move down to organisations that can lose or purloin all your money. And so on.

We have started well. We have agreement on the first two questions.

AH! But the third one, ah! But the third.[116]

Where do we stop?

The third question, like the third Goddess, takes the prize. It is the killer question. It must be asked, and answered, not only where we have agreed that we must regulate (airlines, etc.). It must also be asked of regulation itself. Where does it stop? Where lives, injury and health are no longer involved? Where material loss is no longer involved? What about defamation? What about sacred cows, literal and metaphorical? What about injured feelings? Should regulations prevent you from calling the author a blithering idiot?[117] We make careful note of how difficult it is to determine an objective stop-point.

We also make careful note of the natural tendency of the regulatory empire to grow geometrically. Any regulation must be drafted, published, disseminated, publicised, supported with training, tested and randomly inspected. That gives us drafting, publishing, disseminating, publicising, supporting with training, testing and inspecting departments. Obviously we need staff for these key

116 'AH! But the third one, ah! But the third,
She only raised her beautiful eyes,
Uttered she not one single word,
Yet it was she who bore off the prize.

117 No, they should not. But not too many decibels, please. Remember that there are noise regulations.

tasks. We need personnel, selection, training, staff assessment and staff benefits departments. And canteens, and union offices, and grievance processing, and counselling, and maternity and paternity benefits. And when these resources inevitably fail to remove the paradox of risk from life, obviously we need to increase the resources. Left to itself, the regulatory empire will grow *ad infinitum*. Like the bloody great wheel in a certain song, there is no way of stopping it.[118]

Let us try to tease out some insight into the question of where we should stop. Let us attack the question from two viewpoints.

First let us take a test case, airlines. What should be our stop-points?

CASE STUDY 14.1 STOP-POINTS IN AIRLINE REGULATION

Where are the limits, the stop-points on the regulation of airlines?

Perhaps our stop-points should be where we can no longer achieve regulatory control?

This is not helpful. It fixes no limit. In theory we could attach a regulator to each and every member of the airline's staff.

Perhaps where lives, health and serious loss are no longer involved? This suggests that we will not regulate the airline's choice of cosmetics, blankets or menus. Menus? Might not airlines poison passengers? Yes, they might.[119] Menus therefore need some form of regulation:

– At the least, the ability to take enforcement action.

118 Let us hope that regulation does not have the same ending as the song; it is rather rude.

119 Yes, they do. But their food keeps you slim. Because you eat little of it. All my trousers have shrunk since I left British Airways.

– At the most, the attachment of a regulator to each and every action in the catering process.

This is helpful. *It defines the range of effective regulation.*

– *At the minimum involvement end*, we must have legal sanctions. Without them we cannot regulate, we can only ask. Sanctions are the minimum stop-point.
– *At the maximum involvement end*, our supervision of every action is the maximum stop-point. Why? Because the option beyond that is for us to undertake every action. For this we would need the expertise to carry out the operation. We would be well on the way to becoming the airline operator. And responsibility is passing to us. No thanks. We will stop at supervision.

Traditionally, in Britain the civil service used to work near to the minimum involvement end of the range. It set standards and objectives. It left industry to manage the means of their achievement. When HMG sought to manage – rather than direct – aviation security, its management ineptitude was all too obvious.

Now for our second viewpoint.

At what level of triviality of activity should regulation stop?

The practical test of the point at which regulation becomes impossible does not provide a stop-point. If we go on throwing resources at it, we never reach the point at which regulation becomes impossible.

Nor would a cost-benefit analysis of regulation provide an objective stop-point. Why? Because the quantification of trivial losses (injured feelings, minor inconvenience) in financial or other units is not an objective exercise.

There is obviously a minimum level of risk of injury below which we should not regulate, but how do we decide it?

We get off to a bad start with the realisation that the normal definition of risk - magnitude of loss multiplied by probability – is of limited value, and not only because of the difficulties of meaningful quantification. At one end of the scale, where the magnitude is very high, loss of life, however low its probability we may wish to regulate it. But we cannot. Take a simple example. If one child drowns whilst learning to swim, should we let short-sighted idiots ban teaching children to swim? Stupid question, but there you have the principle. Risk-taking is integral with the learning process. No risk, no learning. At the other end of the scale where probability of occurrence is very high, the magnitude of the loss will decrease to some point at which it is trivial, where it does not merit regulation. Let children play rugby, some will bang their noses. Ban rugby. Let children run about, some will fall over and graze their knees. Ban running about. Let children play cricket, some will be useless and suffer humiliation. Ban cricket. *De minimis non curat* sensible people.[120] Risks are an ingredient essential to the character of life, colourful stitches in its rich tapestry. Ban risk, ban fun. And there you have the agenda of *les miserables*, the Bringers of Universal Misery. These mental retards believe that the experience of anguish is the true purpose of life; to this belief, fun is blasphemy; *ergo* ban fun. Vote Malik and ban the BUMs.

But who decides the point at which regulation is not merited? There is a simple answer. Us, the end-users. Not the regulators. We have already seen that, the more the regulations the bigger the regulatory empire, and the bigger the cut the regulatory hierarchy will arrange for itself. The hierarchy has vested interests, and should therefore be disqualified from the decision-making process. Here we have the first requirement of an objective means of drawing the bottom line on regulation. The second requirement is to assess past incidences and future probabilities. The third is to think sufficiently about the built-in-buts of any regulation: does it prevent learning, does it justify its

120 Sensible people ignore trivialities

costs? I offer specific answers to the plague of regulations (Show Need for Regulation), and specific answers to the plague of regulators (Constrain the Hierarchy), both at the end of this chapter.

LIKE TOPSY, REGULATIONS JUST GROWED

Recently I tidied out my attic. I collected two cricket bats, two squash rackets and six badminton rackets. These I felt would be well-received by local children. I took them to the nearest Sue Ryder charity shop. A pleasant elderly lady said thank you very much, but I am afraid that health and safety regulations do not allow us to accept sports equipment, don'tlookatmelikethat. I took the bats and rackets home and I rang elfansafety. A pleasant young lady made enquiries and said that she could not find any such regulation. If she was correct, it means that Sue Ryder, probably unsure of the exact letter of the regulations, had decided to err on the safe side. We had a case of innocent regulation creep, of shifting sands.

A few years ago I decided to replace a 30-year-old wood burning stove. I was informed that new regulations applied to chimneys. I invited a scheme-accredited tradesman to inspect the chimney and to give me an estimate. He informed me that regulations demanded that I line my chimney top to bottom, £2500. Also that health and safety demanded that he erect scaffolding to the top of the chimney, £2000. Plus incidentals, plus VAT, call it £6000. I pointed out that the chimney was constructed of stone, 4 feet thick, and had been burning without known hazard for 400 years. He said did I want to wake up one morning to find my family dead of poisonous fumes? I said that the chimney was on an outside wall, did not pass next to a bedroom, that wood smoke was not odourless carbon monoxide, that our noses worked, that we would smell, and see, wood smoke long before it emerged in quantities sufficient to poison us. And, as an afterthought, that we knew that the chimney was not leaking because neither sight nor smell of wood smoke had emerged from its 4 foot thick walls in the thirty-seven years we had lived here. He said you will be breaking the law, £6000. I smelled a rat.

I made a large cup of coffee and picked up the phone. I telephoned HETAS, the Heating Equipment Testing and Approval Scheme, funded by purveyors of chimney lining, etc., (wouldn't you know it?), and the Solid Fuel Advisory Service. I was informed that the regulations were issued by the Office of the Deputy Prime Minister. I rang that department. They advised me to download the appropriate schedule of the building regulations. Umpteen pages of thick verbal sludge, so impenetrable it could have been written by the deputy Prime Minister himself.[121] I made more and stronger coffee. I waded into the sludge. I could find nothing that applied to my project. I rang the same three organisations, finding a great number of people who did not understand the regulations. Finally I found someone who did. I spoke to a very bright, very helpful, very outspoken gentleman. I asked him whether he would confirm the information imparted by a scheme-accredited tradesman that, if I did not line my chimney, I would be in breach of umpteen laws, and that I would not be visited in prison by my family because they would all be dead. He said ****.[122] We had a lengthy and enjoyable conversation.

The regulations had again crept. The scheme-accredited tradesman could not, or did not wish to, understand the many pages of verbal sludge, so he erred on the safe side, coincidentally benefiting himself twofold. He transferred all risks arising from the regulations from himself to me. He also turned a £100 job into a £6000 job. This was a case of not so innocent regulation creep, a shifty shove of the sands.

For the vendors of stoves and of chimney linings, it was a self-inflicted injury. I was so disenchanted that I bought nothing. I kept my old stove. It was an old friend and had never attempted to con me. But, to be on the safe side, I employed a scheme-accredited chimney sweep. He swept the chimney. He then asked 3 hours at £35 per hour,

121 Two comments made by John Prescott, the deputy prime minister, demonstrated his staggering ignorance of the most basic elements of risk management. After a rail accident when he was Minister of Transport, he said that money was not an issue in matters of safety. Then, in supposed defence of Gordon Brown, he said that it wasn't the captain that sank the Titanic – it was an iceberg. John Prescott was deeply hostile to the eleven-plus test, because he had failed it. Many would say that he validated it.

122 Since you ask, his exact words were, 'What a load of balls.'

£105 (plus VAT of course), to sweep the register plate. I have swept register plates. I know how long it takes: 12 minutes with dustpan and brush. £105 for 12 minutes equates to £525 per hour. I asked him to leave the house. I was polite; I was very proud of myself.

How many trusting old ladies have been fed similarly convenient misinterpretations of regulations by chimney sweeps to the benefit of themselves and their business associates? And have believed that they constituted a legal requirement? And been relieved of £6000 in similar circumstances? And paid £525 per hour (plus VAT) for the complex, scheme-accredited, task of coordinating dustpan and brush?

Consider the dangers of all regulations. Consider their legal, compulsory status. Consider the accompanying scheme-accredited smokescreens which obscure or misrepresent their meaning. Consider the costs of conforming to unnecessary or irrelevant or incomprehensible enactments. Consider who pays them. Consider their only beneficiaries. Rogues and regulators.

Quis custodies ipsos custodiet? Who shall regulate the regulators? Put them out with the trash. Do not recycle.

MANY WORDS MAKE LITTLE SENSE

The brain's natural, primary language of thinking is one of ideas and images. It is not one of words. Describe a person in one word: nice, nasty, fat, thin, and you instantly convey a simple clear image, a feeling of that person. Try instead to write a comprehensive description of a nice, nasty, fat or thin person. 'Comprehensive' means that you have to cover everything. Lots and lots of words for your reader to wade through, and which do not then convey a simple clear image of nice, nasty, etc. Instead they constitute an awkward, user-unfriendly check-list of the defining characteristics of nice, nasty, etc.

A well-known academic study is quoted by Drew Westen in his fascinating book *The Political Brain*. The academics asked their subjects to try to remember the 95-word description of an unnamed task. The simple words were not easy to arrange into an image or story, into a coherent whole. But, name the task – it was washing clothes – and the words immediately fall into place. The name conveys the image. The image is the primary tool of understanding, the words secondary, an embellishment.

Supplement this academic study with your own experience of instruction books. Do your read them? Do you understand them? Or do you instead dive straight in and get a bit of hands-on experience? In 1976 I received a welcome salary increase. I bought my wife our first dishwasher. I retired to my study with the instructions. I emerged after 45 minutes and announced that I knew how it worked and that we could start it up. My wife said thank you very much, it is already halfway through its cycle. I know. I am thick and my wife is bright. No argument. But the generalisation holds. Try the instruction books for a 747. Six inches thick. Eight weeks in the classroom, three weeks in the simulator, still fairly baffled. A couple of hands-on flights, and happy as a sandboy.

The insuperable problem with laws and regulations is that they are written in words, not images. It is impossible to write a document on a complex matter which covers everything and which is totally unambiguous. I knew no squadron commander who read a new pilot the many books of regulations and then told him that they defined the behaviour required of him. I knew several who said, 'Behave like a **** and you are off the squadron.' Two words: '****' and 'off', and two images. Message received and understood. 'Any questions?' 'No, sir.' Sweet and simple. And that has never been said of any book of regulations.

The impossibility of writing clearly and briefly and unambiguously is one of the forces behind the unstoppable growth and complexity of the legal and regulatory industries. In turn, complexity underlies the difficulties which regulation presents to the end-user. First, he is unlikely to understand it – see above. Second, if he wishes to challenge it, he must employ someone who supposedly does understand it, barrister, solicitor or consultant. Very expensive. He must involve himself deeply in the formal processes. Very daunting. Very time-consuming.

His case will be heard by a member of the establishment. By someone sympathetic to laws and regulations. Someone whose comfortable living depends upon them. Someone whose imperative is the letter of the law. Someone who will not say that the law or the regulation is an ass, and so are the lawyers or regulators who wrote them, case dismissed, establishment to pay costs and get stuffed. If only.

NEUTRALISING THE ENEMY WITHIN

Why, you wonder, only neutralise the enemy within? Why not eliminate them, or turn them: make them work for us?

Because both are impossible. We need a hierarchy because the alternative is anarchy, and that is worse. We cannot make them work for us for very long, because sands shift. Put someone in power and their imperative shifts from our interests to the demands upon them to make decisions. And to regulate. They enjoy power, and its privileges. Most go native. There are a very few who genuinely seek to pursue our interests. They go nuts. We cannot eliminate the hierarchy or turn them. We know that the first rule of risk management is to devote resources to achieving the possible, not to attempting the impossible. Let us therefore see what is possible.

We have three objectives:

1 To identify the enemy within
2 To prevent, as far as possible, the enemy within from subverting our interests
3 To limit the inexorable growth of regulation and regulators

The bedrock of all our efforts is knowledge: knowledge of who our enemies are, and of what they are doing. From knowledge comes understanding, and from understanding comes informed action. Therefore we will look first at the pursuit of truth.

NEW RELIGION ANNOUNCED. WORSHIP TRUTH HERE

Let us establish a new religion, devoted to honesty and the worship of truth.

First Commandment: *A fatwah on deceit.*

First problem solved. Liars and spinners for the chop.

Second Commandment: *Opposition to First Commandment is blasphemy.*

Third Commandment: *A fatwah on blasphemy.*

All problems solved. All enemies of truth for the chop.

And that's it. Fourteen words, three images. WOTSUP – The Worship Of Truth Simple Universal Principle. A total religion in three Commandments. Guaranteed more peaceful than competing varieties. Nobel Peace Prize for Malik? Peerage? No need to be rude. Young virgins queuing to be saved? Only joking, officer.

While the charity commissioners are considering this proposal, let us look at some interim measures.

First we must define the lines beyond which information is not our business. Easy, information on areas which do not have an injurious effect on us. The obvious example is the matter of personal privacy. The private lives of citizens are no one's business save their own. Information becomes subject to legitimate interest if a citizen is active, i.e. impacts on others, in a way which they would not permit if that information were known to them. Thus information on breaches of law: violence, fraud, bribery, etc., are a legitimate interest. Information on private sexual shenanigans is not unless, for example, those involved seek access to our children.[123]

We would considerably extend the scope of our interest in the case of anyone paid from the taxpayers' purse. We would like honest judges and policemen. We would like politicians and planning officers not to take bribes. We would like the Speaker of the House and politicians not to consider taxpayers' funds to be their treat-box, nor to rebuild their several houses, nor to taxi themselves or their wives around the land at

123 Or a job in a concentration camp.

our expense. We would considerably extend the scope of our interest in the case of those of whom a certain quality is a requirement. Eyesight for pilots and numeracy for accountants are easily quantified and established. Honesty for politicians and lawyers could not be established so easily; we would have to rely upon the detection of dishonesty, that is upon the availability of accurate information.

Our base measure of legitimate interest is potential injurious effect on others.

THE IDENTIFICATION OF THE ENEMY WITHIN

Donald Rumsfeld is best known for his malign role in the 2003 US invasion of Iraq. His remark about the 'unknown unknowns' was widely publicised. Connoisseurs of irony will have delighted that this one perceptive observation drew upon him more ridicule than any of his many previous evils. Unknown unknowns is a very important concept, and central to risk management. And that is the last compliment that I will pay to Mr Rumsfeld.

The most dangerous of our enemies within are those of whom we are unaware. And perhaps the most dangerous of the most dangerous are those who do not accept that they are dangerous. A few ideologues in key positions in public organisations can establish great power over the organisation by simple gearing. Let me offer you an example.

A few years ago I investigated the feasibility of establishing a business based upon my system failure methodology (Chapter 8). I was assisted in my endeavours by, amongst others, a young lady adviser from Business Link. She was frankly brilliant, insightful, knowledgeable, the lot. One day she came to our meeting looking rather wan. I asked if all was well. No, it is not, she said, I received my annual assessment yesterday and it was poor. As her ability was way ahead of the Business Link average, I found this almost impossible to believe. It transpired that she had not met two of her targets, the number of Asians and the number of single women whom she had established in business. My point is not that Asians and single women are unworthy, but that a publicly funded organisation had targets which those who funded it (us) might well have disapproved, had they had the relevant information. Tying staff advancement to clandestine targets is an extremely effective mechanism for gearing a clandestine agenda to a powerful organisation.

There are a few occasions where the agenda is successfully hijacked by the organised left wing. Periodically, a flagrant example comes to light.

> I asked Ken Livingstone, the mayor of London, if he could tell me what the cultural Olympics are, especially as he* will fund them. He rambled on about them being for all the people of different ethnic backgrounds in London. I told him I was still not clear. 'It is my intention that you are not clearer,' he replied.

The deliberate deception of taxpayers constitutes the subversion of the public interest. If at law this is not the criminal misuse of public office, it should be.

With such occasional exceptions, there is little evidence that such clandestine agenda is often the product of left wing organisations. It is generally the product of something altogether more dangerous. Idealism. There is always a thick seam of do-gooders running through the community who do not accept the nature of the world and of the human race, but seek instead to impose on reality their personal fantasy of an ideal world. Whence ideologues who see immigrants as more needy than the indigenous population, and give them priority in the allocation of houses, health facilities, benefits and other of our nation's limited resources. Opposition is condemned as racism, a very effective image of offensive behaviour, and plain wrong. These idealists could never declare their feelings publicly, because they would never be elected. Nor are they eager to debate them with their opponents, because idealism is a belief system and therefore, as we have seen, not open to serious challenge, let alone to demolition. The unkind name given by their detractors to this thick seam of idealists is Groliies, Guardian readers of low intelligence in ethnic skirts.

* that is we

In an old house, damp and the concomitant rot, and death-watch beetles (DWBs), are the main source of serious expense. Death-watch beetles attack not soft wood but hard wood, and not the surface but the core. They eat oak beams from the centre outwards. It may be God who brings church roofs down on the heads of sinners, but He uses death-watch beetles as His agents. Their life cycle is seven years, during which they munch away, undetectable. DWBs are a formidable foe. They are four times bigger than common old woodworm, which is a pushover, easily seen and despatched. In the nation, left wing organisations are the woodworm. Idealists are the DWBs. Beware death-watch groliies baring ideals. It is they who will bring our roof down.

Our first step in identifying the enemy within is to assure the free flow of accurate information about their intentions and conduct. The restriction or falsification of information is the first recourse of those who wish to deceive.

CONSTRAIN THE HIERARCHY

Our second step is to superglue to the hierarchy responsibility, and retribution, for their actions. Four measures will suffice.

First, we should absolutely ban the separation of the hierarchy from the consequences of their decisions.

They should be debarred from taking decisions from the effects of which they are exempted. That would put the boot into MPs' index-linked pensions and exemption from Freedom of Information legislation. A Prime Minister sending his country to war should send his family to the front line.

Second, the utter nonsense of the hierarchy controlling the investigations of allegations against them should be terminated immediately. And all cases which have been self-investigated for the past ten years should be reopened under the control of a High Court judge, preferably a hanging judge.

Third, in the event of adverse outcomes, we should reverse the burden of proof. The hierarchy would have to prove that they had exercised due diligence in their endeavours to avoid the adverse outcome. Failure Path® Analysis would be one suitable methodology. Was every Possible Opportunity taken to Prevent this Loss (Chapter 8)? Failure to prove due diligence would attract appropriate punishment.[124]

A step was taken in this direction in the Aviation and Maritime Security Act (1990) (AMSA). In the 1980s and 1990s, airline ground staff routinely subordinated security to punctuality. They correctly identified that the threat of a terrorist attack on any particular flight was extremely low, whereas the threat to their careers arising from a late departure was very high. After Lockerbie, the new AMSA was drafted. Included, very deliberately, was a clause placing ultimate responsibility for security on the chief executives and boards of airlines. The implications of this clause were helpfully explained to them. The intent of the clause was to build the responsibility for the priority of security into the system. It removed the hierarchy's easy defence that there had been no system failure and therefore no failure of the hierarchy.[125]

Fourth, the hierarchy should forfeit any value added by their period of office. Thus for example, any increase attributable to public service in the value of memoirs above its base, probably precious little, would be returned to the public purse. The principle is the same as preventing criminals profiting from their crimes.

Obviously this principle would also apply to the financial and other industries. The law, far from placing responsibilities on hierarchies, appears to allow them to depart the scene of the disaster which they have wrought with as much of other people's money as they wish. Legalised kleptocracy.

124 In the case of lunatic wars, the emptying of bedpans in servicemen's hospitals by the malefactors, their wives, and their advisers for their lifetimes, or until the 500,000 dead come back to life. Whichever is longer.

125 '*No systematic [sic] failure*' was the defence used by Sir Ian Blair, former Commissioner of the Metropolitan Police, when his officers shot dead the innocent Jean de Menezes at Stockwell (22 July 2005) Given the number of shootings of innocents by the Met in recent decades, the acceptance of this defence is amazing. The subsequent PR, investigation and compensation must have cost taxpayers £millions. No fault of the hierarchy of the system. Of course.

SHOW NEED FOR REGULATION

We have seen the difficulty of constraining the growth of regulation. There is no simple defining limit. Instead we have to devise a small armoury of counter-measures.

First, no regulation would be permitted unless an objective need – lives saved, losses prevented – had been shown and independently verified.

Second, no regulation would be permitted unless a compliance cost analysis (CCA) had been conducted, and independently verified. The government has had CCAs in the past, but they have fallen into disuse, largely because the regulators did not like them. Thereby proving their utility, we would say.

Third, no regulation would be permitted unless a built-in-but analysis (BIBA) had been conducted, and independently verified.

Fourth, no regulation would be permitted unless as shifting sands analysis (SSA) had been conducted, and independently verified.

Fifth, would-be regulators would have to obtain their funding from the elected government. Thus they would have to persuade the Treasury of the merits of further regulation. The raising of funds from intermediary beneficiaries would be forbidden, otherwise we would all have chimneys of platinum, and the installers' homes would be paved with gold.

Sixth, each and every regulation would come with a midnight clause. It would automatically expire unless it periodically and successfully renegotiated its way through the first five counter-measures.

SUMMARY – THE ENEMIES WITHIN

1 Our leaders, our regulators and death-watch groliies are the enemy within
2 The free flow of accurate information is the key to controlling the enemy within
3 The penalty for withholding information, or for supplying inaccurate information, should be declared a capital crime. Penalty:

something with boiling oil in it and, let us learn from the Chinese, those to be boiled to pay for the oil

4 Our leaders and the regulators should be subject to their own decisions
5 The burden of proof should be upon our leaders and the regulators to show due diligence
6 Whistleblowers should be celebrated in the Honours List
7 A new religion, WOTSUP – Worship Of Truth Simple Universal Principle – declares a *fatwah* on liars, spinners and deceivers. WOTSUP guarantees more peace than competing varieties, and less Commandments (only three). Also no collections. Join now

CHAPTER 15

SAFETY IN SIGHT

FOREWARNED IS FOREARMED

We have been trying to advance our understanding of risk. It is not a mysterious subject. It is not a territory reserved for experts. It is the territory in which we all live. The better we understand risk, the better we can manage our lives.

We have examined some very helpful ideas put forward by experts. We have developed some ideas of our own, not for our philosophical amusement, but for their practical value in informing our lives. For that reason, we have continually tested them against real life, against well-documented accidents and events. Experts told us that accidents are normal in certain situations. We disagreed. We found that accidents are normal in all situations.

We have followed our own advice. We placed risk in a context which provides us with useful insights. We placed it in the same context as security, that is, in the context of conflict with antagonists. This reflects our common, frequent feeling that the world is against us. We have borrowed from the ancients the concept of all-seeing, all powerful Gods, a pretty difficult bunch of (mostly) chaps. They represent the nature of the world. They are trying to make our activities go wrong; we are trying to make them go right. Our best chance lies in understanding what is going on.

For that reason, we have followed our own advice again. In Chapter 2 we identified the forces arraigned against us, and their order of battle. They start with self-inflicted injuries: ignorance of rules of battle, ignoring the rules of battle, not knowing or not dealing with the forces arraigned against us; the normal causes of which, you will remember, are complacency or hubris or insufficiency of thinking. Further human fallibility is encapsulated in Murphy's two Laws. Nature has its own formidable forces: Sod's Law, options forgone both of objectives and activities, built-in-buts, double agents and shifting sands. This model helps us to understand what is going on.

We looked at what we can learn from previous accidents. We were seriously disappointed to discover that those in positions of power continually failed to heed obvious warnings, or to learn obvious lessons, or both.

We looked at the potential of accident investigation to drive the progress of accident prevention. We were seriously disappointed once again.

We found that most accidents could be attributed to an underlying failure of decision-making. We saw that a decision rests on a stool with three legs: information, intelligence and integrity. We observed that most failures were of integrity.

We revisited the concept of the cause of an accident. We concluded that a new definition of cause was needed: *the failure to take the opportunity to prevent an accident.* We saw that this would be of great value in investigation and in law. It is incorporated in a new methodology, Failure Path® Analysis.

Failure Path® Analysis provides an objective basis for the determination of due diligence. It is thus a means of placing accident prevention uppermost in the minds of the hierarchies of the organisations which conduct or oversee potentially dangerous activities. It would have a very considerable and positive effect on their interest in accident prevention.

We were obliged to take a careful look at decision-making. Decision-making informs action, and therefore directs our path through life. We came to the conclusion that Man, supposedly the rational animal, is rational only in matters superficial. In his depths, and in final analysis, he is ruled by his feelings.

We evolved the Garbage Can Theory of decision-making into the Dolly Mixture Selection Box Theory. From the three compartments of this selection box we extract in turn the factors in the external situation, the mental tools, and the decision-making methods appropriate to the problem before us. We were not pleased to find that the Gods can always rearrange the external situation, and thereby undo our best endeavours. But forewarned is forearmed.

Finally, we looked at the very many, very serious, risks to which we are subjected but over which we have little control. We looked at the establishments and hierarchies who manage them for us, and at their management of those risks. We did not much like either the managers or their management.

We started the book with the question: what's going on? We have tried to find an answer. This has yielded at least one major success. We have identified the key ingredient in our search for salvation. Understanding. For understanding we need, not more information, but better information. The fundamental need is for truth, for a free and uncorrupted flow of information.

In this, our final chapter, we will fill in a few small cracks in our understanding. We will summarise our personal risk programme. We will propose a national risk programme. And we will drink a libation to our readers. And another to the Gods.

THE GOLDEN RULE

We know that our former Chancellor thought that our gold was his, and with characteristic prudence sold it all at rock bottom prices. Fortunately, he did not get his hands on our Golden Rule of Risk Management. We still have it. Here it is. **Look after yourself.**

Because, if you do not, you go the same way as our gold. Up the Swanee. No matter that the accident was not your fault, that some little twerp was joy-riding, or on the wrong side of the road, or drunk. He will get three months probation, counselling, and life on benefits. You too will get life. Underground. For ever. Therefore look after yourself. If you allow yourself to be a victim, it is you who will be the real loser.

It is not only on the roads that life is a self-defence operation. We are surrounded by predators and incompetents everywhere. Therefore life is a self-defence operation everywhere. We know from our study of external situations that our ability to control its factors rapidly diminishes to nil, as their distance from us increases. However, the better we understand them, the better we can defend ourselves against their effects. Understanding is again the key to our risk management. The acceptance of victimhood is the self-pitying attitude of the professional loser. Even more pitiful is their belief that their defence is the responsibility of someone else. Losers they are and losers they surely will be. No great loss; do not be too sympathetic. To ignore a rule of battle is a self-inflicted injury. And when it is the Golden Rule, the injury may be fatal.

Oi, you say, just a minute, what about the superbugs running, sorry, ruining, our health service, what about iffy airlines, and groundless wars? What, you ask, am I meant to do about them? You ask a very fair question. Here are the answers, as far as they go. Remember the Golden Rule. Do your best. Make sure that you are well-informed about risk. Kick up a fuss. Make sure that you are well-informed about the competence of hierarchies – never, ever, take their word for it. Watch them like a hawk. Kick them out if they do not fulfil their responsibilities. You can only do your best, and keep on good terms with the Gods.

THE SEVEN ESSENTIAL STEPS IN THE ANALYSIS OF RISK

We can use our own risk model to analyse our risk. There are seven essential steps.

The first three steps and the seventh step are labelled *strategic* because, if misconceived, they have great capability of bringing failure to the whole operation. The intermediate steps are labelled *tactical* because they have less capability of bringing failure to the whole operation. Their failure is likely to be confined to a part of it.

THE SEVEN ESSENTIAL STEPS IN THE ANALYSIS OF RISK

Essential Strategic Step 1 Options of Different Objectives Forgone
– Consider whether we have chosen the right objective. Consider what other options (perhaps better) we have thereby forgone.

Essential Strategic Step 2 Options of Different Activities Forgone
– Consider whether we have chosen the right activity. Consider what other activities (by which we could pursue our chosen objective) we have thereby forgone.

Essential Strategic Step 3 External Situation
– Analyse external situation (see Example 15.1 below).

Essential Tactical Step 4 Double Agents
– Identify potential double agents, safety measures protecting our operation upon which we may be relying too much, and which may let us down.

Essential Tactical Step 5 Built-in-Buts
– Identify the inherent weaknesses of each element of our operation. Make provisions to deal with their effects.

Essential Tactical Step 6 Shifting Sands
– Identify possible shifting sands. Plan our response.

Essential Strategic Step 7 Overall Picture
– Consider our system of operation in all relevant contexts.
– Review all aspects of it.

1 Are there any cracks in it? Have we overlooked or ignored any rules, possibilities, or factors?
2 Overall, is our policy consistent? Is it free from internal contradictions?
3 Overall, is our policy coherent? Will those involved understand it and be in a position to implement it without introducing internal contradictions?

Essential Strategic Steps 1 and 2 Objectives and Activities Forgone

First, choose the right objective. Choose the right tree.
Second, choose the right activity. Choose the right ladder.
Put right ladder against right tree.
Try to get it right.
Need more be said?

Essential Strategic Step 3 External Situation

We have suggested that there is no law of unexpected consequences. Obviously you could be picky and argue that one cannot think of everything. You would be right because, as we know, the Gods may at any time choose to interfere in the external situation. To amuse themselves, they may shift the sands, change important factors, or introduce new killer factors This is not unexpected. It is to be expected. Therefore there is no law of unexpected consequences. But there is insufficient thinking; this opens the gates to our activity. The Gods will notice and may disapprove. If the gates are opened very wide, they will deem it complacency. Then they most certainly will disapprove, and they will come swanning in, bent on mischief.

Therefore our thinking must be sufficient. We must consider all aspects of the external situation. We must view it in all appropriate contexts. We must identify all relevant, important and killer factors. Then identify the layer beneath them, the factors important to the top layer. Repeat until satisfied that the scope for the unexpected is minimal. Build in a safety factor or a contingency plan, so that this minimal unexpected cannot zap us if it does appear. If the external situation is complicated, map it out, or make a pathway-style analysis of each important factor and its underlying factors. Here is an example.

CASE STUDY 15.1 ANALYSIS OF IMPORTANT FACTORS – TEA FOR TWO

Important Factors – Tea for Two Kettle, Water, Tea, Milk, Cups, Teapot, Sofa, Champagne, Music. Guest, Transport for Guest.

Example Important Factor Kettle

Important to Kettle Serviceability of Kettle, Electricity Supply

First Line of Self Defence Test Kettle, Test Electricity

First Back-up Plan Spare Kettle, Camping Gas Heater

Second Line of Self Defence Test Spare Kettle, Test Camping Gas Heater

Second Back-up Plan Spare Champagne

Why, perhaps you ask, champagne? Because our subject is risk management, and the Second Golden Rule of risk management is KISS – Keep It Simple, Stupid. Hence champagne – no corkscrew needed, no corkscrew to lose. Simple.

Did you read the case study carefully? Are you satisfied that no important items were omitted from our analysis? I hope not.

There is one deliberate mistake, the only (I hope) dirty trick in the book. Matches, to light the stove in our first back-up plan. I am sure that you spotted their omission. Would it have undone us? No. The champagne would have saved the day.

The successive differentiation of important factors into their further underlying layers of important factors is essential to the analysis of the external situation. It is essential to expecting, and thus eliminating, the unexpected. Without sufficient thinking, Losers Is Us. With sufficient thinking, Winners Is Us (*deo volante*– by the kind permission of the Gods).

Obviously, if instead of climbing a tree or making tea, our activity were making motor cars or running a railway, or opening a new terminal at Heathrow, the analysis would involve rather more factors. In the case of our railways or of opening a new terminal at Heathrow, rather too many for those responsible. But the principle holds good, identify all important factors. Build them into your plan.

Essential Tactical Steps 4 and 5 Double Agents and Built-in-buts

We explored double agents and built-in-buts carefully in Chapter 2. You may have identified their appearance in some of the real-life examples throughout the book. We need no further detailed discussion of them.

Essential Tactical Step 6 Shifting Sands

Anything and everything in our plan may be subject to shifting. For that reason, the consideration of shifting sands cannot take place until all elements in the plan are known. Therefore this is our penultimate step. True, if at this late stage we identify any such shifts or the possibility thereof, it may be a bit of a bummer.[126] But much less of a bummer than total failure.

126 And perhaps a case of *Back to the Drawing, Bored Hoskins.*

Essential Strategic Step 7 Overall Picture

The first task of the Man with the Overall Picture, if you can ever find him, is to look for gaps in our thinking and in our system of operation. He is our crack detector. Cracks are openings for the ever-watchful Gods to put their oar in. Cracks are openings through which our antagonists can attack us.[127]

The second task of the man with the overall picture is to ensure that both our policy and our operation are coherent and consistent, and that they successfully navigate the inevitable dilemmas (see immediately below).

Therefore Step 7 is strategic. It is there to save us from disaster, particularly from self-inflicted disaster. It must be closely cognisant of the dangers of insufficient thinking. It must identify and close gaps in our thinking and in our operation. It must ensure the sufficiency of our thinking. Do not forget the matches. Do not leave gates open. Do not, as reportedly did the makers of Airbus, use different, incompatible computer systems for different pieces. Join up your thinking if you wish to join up your tea party, your borders, your railways, your airport terminal, or your aeroplane.

DILEMMAS

Protagoras. Who? Another bright ancient.

Protagoras' maxim states that, in every question there are two sides to the argument exactly opposite to one another.[128] Whether this maxim does or does not constitute a universal law, it makes a point which is always worth considering. It is certainly relevant to decision-making and to planning action.

Protogoras' insight, his identification of the ubiquity of dilemma, underlies our concepts of options forgone, built-in-buts and double agents. What tiles should we manufacture (Chapter 10, Case Study

127 Or, in the case of guarding Britain's borders, for illegal immigrants to enter. Our border force is not funded for working at weekends. So, Monday to Friday: nice high fence; Saturday and Sunday: open days, all welcome. Unbelievable. But do not blame the border force; blame the Treasury which will not fund the weekends.

128 Do not be tempted to say that he is wrong. He would say that you only prove him right. Irritatingly clever, the ancients.

10.1)? A specialist range will secure a niche market – *but* it will not secure the safety of a broad market. Spend the next hour sunk into a comfortable chair and we will de-stress – *but* we will not benefit from exercise in the gym.

Let us take an example of dilemmas in action.

You live in a country which has an active terrorist group seeking independence for their region. You have just been appointed the nation's counter-terrorist supremo. You have to construct the nation's strategy. Please answer the questions in Case Study 15.2, p 308.

There are of course, many more than six dilemmas.

A root problem is the number of different audiences of whom you must be aware. Soft treatment will infuriate those who detest terrorists. Hard treatment will upset those who have some sympathy with the terrorists, and may cause them to increase their support. Most countries will be concerned only to avoid involvement; they will not even be supportive. They will buy their own immunity from attack by doing financial or facilitation deals with the terrorists. A few countries with experience of terrorism may be supportive.

You, our supremo, must navigate us through these dilemmas. You must ensure that, as far as possible, internal contradictions are eliminated. And where that is not possible, that they are constructively managed. Any policy is doomed if it is inconsistent, if it contains unmanaged internal contradictions. It will self-destruct.

Are you sure that you want the job? [129]

This is all rather doleful. I apologise.

However, our example is a clear, if extreme, illustration of a system of activity under permanent threat from antagonists. Each of us is similarly under permanent threat from the nature of the world. Our present point is the need for each of us to take responsibility for

129 If you wish me to get off the fence and give my answer, here it is. The key is increasingly and unceasingly to marginalise the terrorists, so they find themselves pariahs, outcasts without support or sympathy. Therefore the key battle is for the public mind. Also, kick the table over, change any laws which obstruct your purpose. Separate terrorism totally and unequivocally from everything else. Redefine treason. Punish minor breaches with immediate withdrawal of benefits, greater breaches with permanent loss of citizenship and of rights of residence. Extend the full panoply of human rights only to those who behave like human beings.

CASE STUDY 15.2
DILEMMAS – COUNTER-TERRORISM

1 *Should we consider ceding some of the terrorists' demands?*
This might bring peace
But It would demonstrate the effectiveness of terrorism, and encourage its further use

2 *Should we negotiate with the terrorists?*
This might bring peace
But It would give them political legitimacy

3 *Should we make a massive response?*
This might be effective in reducing the threat
But It would imply to the public and the terrorists that the terrorists constituted a real threat and not, as we had affirmed, an irrelevant pinprick

4 *Should we treat terrorists as common criminals, not as political activists?*
This would remove the cloak of political legitimacy for their actions
But It would make it difficult to enact special, severe, legislation

5 *Should we make widespread request for public cooperation?*
This might yield useful intelligence
But It would give publicity to the terrorists which they would welcome

6 *Should we implement massive security at airports and on the underground?*
This might prevent successful attack
But It would massively inconvenience the public

the management of his/her own risks. Each must decide their objectives and identify the important factors in their situation. Each must devise a plan which is internally consistent, and which navigates its way through the obstacle course of dilemmas. Each must then manage the successful implementation of the plan. Our enemies are at the gate, always trying to get in. It is up to us join up our thinking and our actions and keep the bastards out.

THE REAL RISKS OF REGULATION AND FOXTROT INDIAS[130]

We have already seen that regulation, by its nature, cannot stop growing. Therefore we must stop it. We have reviewed a number of counter-measures: tight budget, show need, midnight clause, *et al.* (Chapter 14).

Let us look at regulations' three most cancerous properties.

First, it is impossible, as we have seen, to write simple, clear and unambiguous regulations. It is not only that the parliamentary draughtsmen are not up to it; it is that language is inherently ambiguous. Why are lawyers so rich? Largely because there are always different interpretations of the letter of the law. Thoughtful parliamentarians will point to obvious possible misapplications. The cretin minister of the day will always reply that the law was never intended to be/would never be used like that. Rubbish. *If a regulation can be misinterpreted or misused, some idiot will do so* (Malik's First Rule of Regulation)[131]. Legislation intended solely to address the extreme offence of terrorism is applied right, left and centre. It is applied to T-shirts with amusing captions; it is applied to an 82-year-old heckler at a New Labour Conference.

Do not overlook the evil undercurrent of most regulation. Detailed regulation is primarily to the benefit of the regulators (them) and to the detriment of the regulated (us). We are bound by the regulations; therefore we can be held responsible for any breach. But here's the nice bit. The regulations, as we know, will impede the execution of the task. They will not cover the real-life situation, either on account of its complexity or because the sands will shift. Therefore, to get the

130 The term 'Foxtrot Indias' is explained in these next paragraphs.
131 Malik's Four Rules of Regulation are listed in the Appendix to this chapter.

job done, we must ignore some of the regulations. And if anything goes wrong, guess what? You're fired. Do you remember the happy days of industrial strife? Do you remember Work to Rule – the meticulous adherence by the workforce to each and every regulation, no telescope to the blind eye, no exercise of discretion? Work to rule very quickly brought everything to a grinding halt. The reality is that, if you want to get the job done, you have to let the regulations go hang. Much better, let's go hang the regulators.[132]

Second, galloping regulation spreads throughout the land the attitude that the state is responsible for managing every risk of our lives. It engenders the belief that the state will hold over us an umbrella which will protect us from everything. Everything ranges from the trivia of jumbo jets dropping on our heads to the massive mental injury from the disappointment that Miss Piggy Tarts contain no pig. It is a big umbrella. It costs us lots. It needs many, many, people to carry it. It keeps lawyers and regulators prosperous, and death-watch groliies deeply self-satisfied. It spreads the expectation that, since the state is responsible for managing our every risk of life, we can sail along thoughtless, with our mind in neutral and our thumb not in useful employment. If we sail into disaster, poor us, we are victims; the system has failed to protect us. Who do we sue?

Third, as if the second were not enough, it engenders the attitude that the state's umbrella absolves us of the need to consider the well-being of our fellow-citizens. Thus, since all risk (bar natural risk) emanates from people, we have a society of people emanating risk, because they are under no obligation to do otherwise.

And the answer – more regulation? Like hell. The answer is a change in social attitudes. Responsibility for the risks which can manage. Acceptance of utterly trivial risks. More consideration for

132 Peter Brothers, distinguished wartime fighter leader, told me about the Atcherley twins. They were wartime RAF pilots, later senior officers, brave, successful and enormously popular. They were outrageously and rightly scornful of RAF regulations. Their attitude was: make your decision; then consult the book; if the book permits your decision, cite the book; if the book does not permit it, observe that the book has always been known to be rubbish, and sling it in the bin. Whichever the case, carry out your decision. Atcherleys, Gentlemen, we need you now as never before.

fellow-citizens. It will lead to more mutual trust, and to the change in social attitudes which we need. Massive pressure on our leaders to manage the risks which are their responsibility. One of many reasons for the importance of trust in our leaders. They are our role models. God save us.

Let me end this discussion with a small confession. Whether it was the manhandling of an 82-year-old heckler deemed a terrorist, Mr Prescott opposing good schools because everyone would wish to go there, Mr Blunkett whining that he was working-class lad undone by a millionairess (albeit accurate in one respect), Miss Piggy Tarts, the ban on the exceedingly dangerous cakes made by members of the Women's Institute, or one of several hundred other recent possibilities, I cannot remember. One of them prompted me to express my total exasperation in regrettably blunt terms. 'F***ing idiots'. I dislike profanity, and therefore sought the advice of a moderate and urbane gentleman. Could he, I asked, suggest another phrase to describe those involved? No, he said, you could not possibly put it better.

This leaves me to try very hard not to use the phrase. It is not easy. There are Foxtrot Indias everywhere.

WHAT DO WE DO ABOUT THE HIERARCHY?

Let us pass delicately over those answers, many and excellent, which might bring a police-car to the door.

We need, as we have seen, to tie hierarchies into the consequences of their actions. Hierarchies, whether Prime Ministers, politicians, commissioners of police, MEPs, MPs, company directors, investment bankers, non-domiciles, or nice simple multi-billionaires, do not live on the same planet as us, the common people. They live on a planet on which special laws apply. A planet on which immunity from retribution is the law. Where the UN blocks inquiry into its corrupt operations, where the annual accounts of the much-admired European Commission have been rejected by its auditors for twelve years, where MEPs and MPs block inquiries into their expense claims, where involving your country in a lunatic war makes you a multi-millionaire, and where leading a public company to destruction earns you a £massive golden goodbye, presumably to spend on hankies to dry your eyes.

Obviously, given our views on regulation, we are not proposing new laws. We are proposing that hierarchies are subject to existing laws, like everybody else. This means giving oversight bodies real teeth, fixed not removable. It means that the establishment no longer oversees itself, and that those under investigation do not do the investigating. Rocket science? Hardly. It is a long-established principle of investigation. Who knows, one day it might spread to MoD and the RAF and the House of Commons. A cheering thought, but don't hold your breath.

Obviously those who are spending our money will be subject to more stringent examination. When you come to consider it, that means each and every hierarchy because, in the last analysis, everything belongs to us: politicians, the civil service, companies, banks, the lot. So when they hand themselves their money, they aren't. They are handing themselves our money. We carry the risks of their management. And we pay for it. In both senses.

SAVED BY OUR NEW RELIGION

Our proposal for a new religion, the Worship of Truth Simple Universal Principle, or WOTSUP, is with the charity commissioners. Surely they will not turn us down. They would not wish to be seen to be opponents of truth and honesty.

Our quest for truth in the matter of risk management started with the question: What's going on? It is not a quest to which there is an end. It is a quest in which continuous progress can be made. For this we need stepping-stones of truthful information. It is a quest in which spin or inaccurate information or deceit provides momentary false support and then dumps us in the youknowwhat.

Therefore we must construct a bodyguard around truth. Following allegedly deliberately false rumours in the money markets, there was a discussion of the need for a whistleblowers' charter in the money markets. The need is not for a whistleblowers' charter confined to the money markets. The need is for a whistleblowers' charter and for freedom of information everywhere, particularly where the tentacles of government touch our interests and touch our pockets. Like I said, everywhere.

Risk itself is real, but that reality may easily be obscured by false perceptions. For that reason, truth must rule. Those who peddle porkies should be ground up into fillings for Miss Piggy Tarts. We need to know which countries threaten us and which do not, which airlines are safe, and which less so, whether roads are made safer by more road signs or by fewer.[133]

Truth has two enemies: vested interests and sacred cows (metaphor). The former try hard to develop into the latter. The casting of doubts on sacred cows is of course blasphemy. Very convenient to those who live by milking them. Don't fall for it.

WOTSUP is a very tolerant religion, but its tolerance does not extend to vested interests or to (metaphorical) sacred cows. Nor, if you are a worshipper of truth, should yours.

CASE STUDY 15.3 THEREBY HANGS A TALE OF A SACRED COW

In 1972 a newly qualified psychologist, David Rosenhan, persuaded eight of his friends to present themselves at psychiatric emergency rooms across America, and to say, 'I am hearing a voice. It is saying thud.' All eight, unkempt after five days without proper ablution, were admitted to hospital. Once in, they all said that the voice had disappeared and that they felt fine. They were kept in hospital for periods of between seven and fifty-two days.

133 The answer is, in many circumstances, roads are made safer by fewer signs.

In a number of cases, the other patients seemed to know that the experimenters were normal. One said to Rosenhan, 'You're not crazy. You're a journalist or a professor'.

Rosenhan published an account of his experiment in the prestigious journal *Science.* Some years later, the author of the book from which this account is taken had a telephone conversation with a 'most prominent psychiatrist'. When told of some terrible mishaps which had befallen Rosenhan (death of wife from cancer, injury of daughter in a car crash, and several strokes himself, and paralysed), this most prominent psychiatrist replied, 'That's what you get for conducting such an inquiry'. Nice chap. Must consult him sometime about my mental balance.

Few vested interests are as obnoxious as that prominent psychiatrist in their expression of self-regard. However, most, in my experience, are aggressive and threatening, and routinely mendacious. In earlier days I was editor of the British airline pilots' journal. I have a collection of very shirty letters in response to unfavourable, but absolutely accurate, comments – from an airport I described as somewhat dangerous (Kai Tak, the old Hong Kong airport; brilliant fun for pilots; nonetheless somewhat dangerous); from a now respectable airline (shan't tell you) one of whose pilots was involved in an accident after he had worked dangerously long hours; from the manufacturer of an aircraft, the tail of which fell off, etc. Attempts to deny or suppress unfavourable facts are the norm.

Recently Canada was forced to withdraw Israel and the USA from its list of countries practising torture. If the publication of the unpalatable facts of unpalatable behaviour damages an aspired image, so be it. The First Commandment of WOTSUP says publish or be damned. To recover the aspired image, stop the unpalatable behaviour. Easy.

There is only one occasion on which it is acceptable to skate over unpalatable facts. It is constructive during the active negotiation of the resolution of an issue. Conflicts will never be resolved if negotiators insist on cataloguing the atrocities of the other side. This they should avoid, instead proceeding directly to solid common ground for negotiation. In all circumstances other than committed negotiation, truth must rule supreme. No other waivers from WOTSUP.[134]

There is a long list of sacred cows who regard any questioning of their conduct as blasphemy. A Good Cause is very often good camouflage for their activities. The agenda, and thus the funds, of a good cause can easily be hijacked by those of the death-watch groliies' persuasion for purposes which would not be supported if they were known. Hence the charge of blasphemy, its intention to deny the free flow of accurate information. Therefore enquire precisely *cui bono*, who benefits from national fund raising, for example Comic Relief. Enquire what benefits the administrators or organisers extract from the charitable funds. Enquire precisely where the remainder goes. Examine the basis of green causes – is it based on facts, or is it based on beliefs recycled as facts? Enquire about the overall impact of their propositions. Examine the sufficiency of their thinking. Do not expect to be popular.

Enquire how much of our money the National Lottery contributed to the Battle of Britain Memorial. The answer? None. The Battle of Britain is not of sufficient general interest. Long pause, while you take that in. Where would Great Britain now be without the Battle of Britain?[135] The National Lottery was established to support four broad areas: arts, sports, heritage and charitable causes. Many regard the Battle of Britain as an unsurpassed episode of greatness in our heritage. But of course, the death-watch groliies of the National Lottery know best. Cross-gender citizens and edible guinea-pigs in Peru are clearly more worthy of our money than the nation's memory of the bravest of our brave. Silly me.[136]

Also enquire how much of our money HMG contributed to the Battle of Britain Memorial. Absolutely nothing. Sweet Fanny Herself.

134 Except when your life is in danger, e.g. In responding to a lady's question on the elegance of her new hat, or on the slenderness of her seat.

135 *Nicht eine schwere Frage.*

136 Had we lost the Battle of Britain, we would not be plagued by death-watch groliies. Hmm.

Him Prudence, then chancellor, presently Prime Minister, now urges us to celebrate our Britishness.[137]

The Battle of Britain generation reaffirmed the Great in Great Britain. Our generation of death-watch groliies and Foxtrot India politicians has erased it. So let us now forget the Spitfire, an inappropriate, military, symbol. In its stead let us, and future generations of Britons, hang pictures of edible guinea-pigs on our walls. Let us gaze at them in envy and admiration, and discuss whether the purity of the Mk 1 Guinea-Pig was to be preferred to the power of the genetically-modified Mk 9. And let Remembrance Day be a celebration of our cross-gender citizens. And don't ask what Poppy Day will be called, you will upset people.

Is this relevant to the subject of risk? No. Unless you believe that social cohesion and mutual civility rests on loyalty and commitment to the country. And that your loyalty and commitment rests on your admiration for its heritage of great heroes, its people, and its values. That it rests on the ashes of our fathers and the temples of our Gods.

But if you believe that, you are out of touch. Like me.

YOUR PERSONAL RISK PROGRAMME

Each of us has his/her own personal risk setting.

If you do not feel alive unless your life is hanging by a thread, then you will feel fulfilled only if you are hanging off the Eiger on a string in a blizzard, or if you are cocking close-range snooks at polar bears, or if you are undercover in the Mafia. Go ahead. Enjoy.

If you are preoccupied with safety, become an airline pilot and enjoy combating the management, the weather, the US Navy, and sundry other threats. Then write books on risk.

If you are exercised only by the thrill and challenge of large cream cakes, find a good psychologist to help you to get a life.

137 To suggest that Machanical M'Doom (Chapter 10) is trying to lose his Mac would be cynical. Or cynical realism.

Each one of us must take the advice of that clever lot, the ancients: *Gnothi suetoni* – know thyself. Find out where you are sited on the risk scale; live there, that is where you will be happy.

Next, construct your own definition of extreme adverse events. Then avoid them. If you wish to take risks, do so thoughtfully, not thoughtlessly or ignorantly. Make your list of pros and cons and look hard at the cons. Treat with scorn any precise figures of probability. A one-in-a-million chance of whatever has a one-in-umpteen-million chance of being honest and accurate. Remember Sod's Law. If it is a one-in-a-million chance of winning the lottery, you won't. If it is a one in a million chance of losing your money or your life, you probably will; think Russian roulette with a two-chamber revolver.

It is the nature of the world that happiness is long and slow to construct, but unhappiness is a hair-trigger away. Ten thousand OKs are undone in an instant by one O Sh#t. Hence our need to avoid the one extreme adverse event.

Our actions are directed by our decisions. Remember that you take a decision, not when you take the decision, but when you make the subjective judgement of the basis upon which the decision will be taken. Decisions rest on a three-legged stool. The leg most likely to be wonky is integrity, the integrity with which the decision-making process is tied to the right imperative. Dive into the second compartment of the Dolly Mixture Selection Box, into your mind, into your mental processes. What is driving this decision? Is it the overt imperative of rationality? Or is it by any chance a covert imperative: personal vanity, personal enmity, or other deep-seated, perhaps unworthy, need? Teenagers are driven by the need to assert their personal self-esteem or their independence from their parents. Hence their moronic decisions and their death-rate on the roads. The dimmer types of men are driven by the need to appear macho. Hence their arguments on any subject in which neither will concede ground. In reality, such dick-length contests are not about proving who has the biggest dick. They are about proving who is the biggest dick.

The three concepts – imperative, integrity and the subjective basis of the decision – are closely related and often coincide. Therefore always keep an eye on your imperatives. That's a must.[138]

Beware of insufficient thinking. Remember the seven essential steps. Mull over the dilemmas. Keep one eye on the overall picture. Keep the other open for mischief incoming from the Gods.

OUR NATIONAL RISK PROGRAMME

My second son is a large rugby player. When he was twenty, he telephoned me with a plaintive request. 'Can you do something about your mother?' 'No one can do anything about my mother. What's the problem?' Apparently in the local supermarket he had tried to pay for some confectionery which mother had bought for him. Mother had told him to back off. He had failed to back off, and so she had hit him. Mother being a gentle soul, five foot not very much, and ninety-two years old, the blow occasioned no great physical damage to his muscular frame. However, several members of his rugby team witnessed the assault, and were hospitalised with laughter at his expense. For which reason the incident effected great damage to his sensitive soul.

Similarly you may feel that no one can do anything about our hierarchies. Yes, and then again, no. In the short term, all that we can do is to construct defences around ourselves to reduce the risk which they pose. In the longer term, let us deploy a double whammy. The first whammy (however pleasing the thought of Semtex suppositories), is a set of new and seamless laws.

First, to attach firmly to individual malefactors in hierarchies the responsibility for their actions.

Second, to deny to those malefactors access to our money to fund their defence costs.

Third, to place the burden of proof upon the malefactors to show that they had exercised due diligence in their conduct.

138 Joke. OK. Wet joke.

These new laws would be supported with a nationwide campaign against corruption and incompetence. All this would steadily corral hierarchies into an ever-reducing stockade, limiting their scope for self-serving activity and for depositing their mischief upon us. Whistleblowers would be rewarded with a seat in the Lords. There they would continue their good work, whilst keeping a sharp eye on the shysters who had bought their way into the same place with money or other favours.

The second whammy is the adoption of our new national religion, WOTSUP, Worship of Truth Simple Universal Principle. Spin, kleptocracy, declarations of unjust wars, and the purveying of extreme stupidity to be deemed blasphemy and rewarded fittingly. Spinners to drive round the M25 for 25 years. Kleptocrats to be housed with pigs and live on pig-food for ever. Declarers of unjust wars and their families to empty bedpans in servicemen's hospitals for ever and a day. Death-watch groliies to be isolated from the human race, confined to an otherwise uninhabited Scottish island, and left to bore each other to death. Or sent to the Taliban.

SAFETY IS NO ACCIDENT

You may recall our Revisionist Normal Accident Theory. Do not let its heavy name obscure its simple truth. Accidents are normal. Accidents, not safety, are the default mode of life.

Accidents are nature's way. To that extent, nature is our antagonist. Safety is the continuous process of defending ourselves against nature, both against nature's little ways and against our own nature. Safety is no accident.

THE GROWN-UPS BOOK OF RISK – ELFANSAFETY WARNING

I believe the contents of this book to be both true and accurate.

And I urge you not to believe a word of it.

That is, until you have satisfied yourself that it is both true and accurate. Use Popper's test of falsification. Take incidents from your own life. Take newspaper reports of events or of human behaviour. Say to yourself these are not explained by The Grown-Ups' Book of Risk, by the concepts of self-inflicted injury or built-in-buts or double agents or shifting sands or the three-legged decision stool, or by other concepts in the book. If they are explained, raise a glass to the author in his cardboard box. If they are not, try again, this time remembering the concept of Foxtrot Indias. If still they are not explained, raise a glass to yourself for bettering the author.

FACTS RULE. OK?

Risk management is about the avoidance of serious loss.

It is about hard reality, about finding the facts, understanding them and facing them.

The withholders of facts, the spinners of facts, and the peddlers of lies, are serving themselves by doing you a disservice. They may cause you serious loss. That will not bother them. They are the enemies of the successful management of our risks.

The management of the risks of our lives is an unending battle. Enjoy it. To win, we must go into battle facing the facts. Those who go into battle in blinkers will not see what hits them. Therefore keep your eyes open for all-thumbs Murphy and for the ubiquitous lurking Sod. Never think you have got it made. Never ever say that you have got it made. Remember the guidance system fitted to thunderbolts. Hubris-seeking.

Finally, try not to attract the attention of the Gods. If you do, doff your hat, raise a respectful glass to them, and laugh at their jokes. Keep laughing. That way they may take a liking to you. Then your risk management, and your life, will be on the best possible footing.

Good luck. Observe the teachings of WOTSUP. And may the Gods be with you.

APPENDIX TO CHAPTER 15

MALIK'S THREE LAWS OF LIFE

1 All people and organisations will always pursue their own interests, as they see them at the time
2 All people and organisations will always try to get away with whatever they think they can get away with
3 Fool-proofing always produces bigger fools

MALIK'S FOUR RULES OF REGULATION

1 If a Regulation can be misinterpreted or misused, some idiot will do so
2 All Regulations can be misinterpreted or misused
3 No Regulation, however lengthy and detailed, can cover all eventualities
4 The only certain beneficiaries of Regulations are the regulators

Malik's Four Rules of Regulation apply to regulating, legislating, and ninnying and nannying in whatever guise.

MALIK'S FORK

To live, or not to live; that is the question:
Whether tis nobler in this life
To suffer the slings and arrows of outrageous fortune
Or to be bored to bits by death-watch groliies,
And rogered rigid by rampant regulators.

With apologies to the Bard. With absolutely no apologies to others mentioned, and certainly none to Cardinal Morton of Morton's Fork.

This prototype S H One T is the role model for succeeding generations of regulators and hierarchies.[139]

139 Cardinal Morton, Lord Chancellor and money raiser for Henry VII, argued that those who apparently had no money were hiding it. Therefore they could give some to the king. Alternatively, if they obviously did have money, then they could give some to the king. So Britons have been forked by their rulers for over 500 years. We are in good company. *Illegitimi non carborundum.*

COPYRIGHT ACKNOWLEDGEMENTS

I wish to thank all the copyright holders listed below. They have generously, and mostly quickly, given me permission to use their material.

I have made every endeavour to obtain all necessary permissions. If I have inadvertently used copyright material without permission, I apologise and will correct the error at first opportunity. Any such quotations will have been memorable enough to have lodged in my mind, sadly unaccompanied by the appropriate reference.

Air Accidents Investigation Branch

Air Accident Report 3/2003. London: TSO. 2003

Ashgate

James Reason. *Managing the Risks of Organisational Accidents.* Aldershot: Ashgate. 2005

Bloomsbury

Lauren Slater. *Opening Skinner's Box: Great Psychological Experiments of the 20th Century.* London: Bloomsbury. 2004. Extract: *The Man Who Made Fools of the Psychiatrists,* in The Week, 7 February 2004, p 40–1.

Gerd Gigerenzer

Gerd Gigerenzer. *Gut Feelings.* London: Penguin. 2007. Figure 9–2 Fast and Frugal Tree, p 174, after Green L.A., and D.R Meyr. *What alters physicians' decisions to admit to the coronary care unit?* The Journal of Family Practice 45:210–26

Ian Allan Publishing Ltd

Brookes, Andrew. *Disaster in the Air.* Shepperton: Ian Allan/Penguin. 1992

Robert Matthews

Robert Matthews. Tumbling Toast, Murphy's Law and the Fundamental Constants, in *European Journal of Physics* 16 172–176 1995

National Review

Christina Hoff Summers *Summers Storm*, National Review, February 14, 2005. © National Review, Inc., 215 Lexington Avenue, New York, NY 10016. Reprinted by permission

Penguin Group

Gerd Gigerenzer. *Gut Feelings.* London: Penguin. 2007

Richard P Feynman. *What Do You Care What People Think?* London: Penguin. 2007. Reproduced by permission of Penguin Books Ltd

Princeton University Press

Charles Perrow. *Normal Accidents.* Princeton University Press. 1999

Scott Snook. *Friendly Fire.* Princeton University Press. 2000

Scott Sagan. *The Limits of Safety.* Princeton University Press. 1993

Simon and Schuster UK Ltd ©

Andrew Weir. *The Tombstone Imperative.* London: Pocket Book. 2000

Taylor and Francis

John Adams. *Risk.* London: Routledge. 2004

News International

Simon Jenkins. *The rise of imperialism lite is prolonging the Iraqi horror.* Sunday Times, March 16 2008, p 1.16

Richard Brooks. *Biteback.* Sunday Times, March 16 2008, Culture Section, p 14

Border force to shut at weekends. Sunday Times, March 2 2008. p 1.5

Peter Nicholson, letter to *The Times*, reprinted in THE WEEK, 2 Feb 2008, p 21

Sunday Times. *US Hits Panic Button.* Oct 21, 2007. p1.24

Brian Toft

Brian Toft. *External Inquiry into the adverse incident that occurred at Queen's Medical Centre Nottingham 4th January 2001*

NOTES

One of the pleasures of the subjects covered in this book is the number of excellent books which address related areas. A book or paper worth reading is, I submit, one which provides well-written, well-argued, food for thought. All the books and papers mentioned in this book and these notes meet that definition. One or two on the fringe are academic and somewhat harder going; they are for those who are seriously interested in their subject matter. All the others I warmly commend to you. Please remember that you are not obliged to read a book cover to cover. Whizz through it, make note of the interesting bits, then dig into those and go on digging until you have extracted all that is of value to you. When you come to bits which you think tell you nothing, don't hang about, your first impression may well be right. I learned that the hard way when I was studying terrorism. I read a number of lengthy tomes, at the end of which I wondered what I had learned. So I read them again, only to find that they did in fact say nothing, and at considerable length. Adler and Van Doren's *How to Read a Book* is astonishingly good. And, yes, I can guess what you are thinking. Mock ye not, at least until you have read it.

The number at the start of each reference is the related page number in this book.

CHAPTER 1.

13. Robert Matthews. Tumbling Toast, Murphy's Law and the Fundamental Constants, in *European Journal of Physics* 16 172–176 1995

Robert Matthew's paper is both thoughtful and entertaining. We may hope that his incontestable conclusion, that toast obeys the laws of nature, will not put an end to all entertaining discussion of the subject.

16. *The Times*, letter from Peter Nicholson, reprinted in THE WEEK, 2 Feb 2008, p 21.

The humour in the lighter letters to The Times is one of the great delights of England. How long before the groliies take offence and try to ban it?

CHAPTER 2

36. John Adams. *Risk*. London: Routledge. 2004

John Adams is Professor of Geography at University College, London. He is an iconoclast who swings a well-directed boot into the false assumptions which drive so many safety crusades. A great book, full of solid food for thought. Hands up all regulators or groliies who have read it; anyone? I thought so.

38. *RAF Proceedings of a Board of Inquiry into the Accident to Chinook HC2 ZD 576 on 2 June 1994*

The proceedings of this Board of Inquiry were released under John Major's open government initiative.

38. Brian Toft. *External Inquiry into the adverse incident that occurred at Queen's Medical Centre Nottingham 4th January 2001*

Brian Toft is Professor of Health Risk Management at the University of Coventry. He is one of the country's, if not the world's, experts in the investigation of medical accidents.

CHAPTER 3

58. Charles Perrow. *Normal Accidents*. Princeton University Press. 1999

Normal Accidents is the second of two landmark books in accident aetiology (causation); it was originally published in 1984. The first landmark book is Barry Turner's *Man-Made Disasters,* published in 1979. In Perrow's afterword to the 1999 edition of his book he points to a series of accidents which unarguably arose from the failures of large organisations (Bhopal, Chernobyl, *Challenger*, environmental pollution and spills). These established the credentials of Turner's and Perrow's works in mainstream academic thinking. Unfortunately their concepts are still not yet embedded in judicial thinking. This is to be deeply regretted because many hierarchies – ministers, civil servants, the RAF, the NHS –are likely to remain in denial until legal precedent and public outcry oblige them to acknowledge that it is their management responsibility to eliminate situations predisposed to the occurrence of accidents.

63. Both quotations are from Adams (Note on Chapter 2, above)

64. Richard P Feynman. *What Do You Care What People Think?* London: Penguin. 2007, p 216

Richard Feynman was famous as a brilliant theoretical physicist. *What Do You Care What People Think?* shows that he also had his feet firmly on the ground. Much of his public fame came from cutting through the obfuscation and focusing unshakeably on the killer factor in the *Challenger* disaster, the brittle O-rings, recounted in this book. His writing is informative and highly entertaining.

65. Gerd Gigerenzer. *Reckoning with Risk.* London: Allen Lane/ Penguin. 2002

Gigerenzer is the third on my list of three thoroughly good eggs, Adams, Feynman and Gigerenzer, whose books I cite. All three are iconoclasts and speak their minds; all three put the boot deep into half-baked or self-serving thinking; all three are very entertaining to read. In *Reckoning with Risk* Gigerenzer explains some important statistics in wonderfully simple terms. If you have the misfortune to need to understand medical or forensic statistics, read *Reckoning with Risk.* Then make your doctor or lawyer do the same.

66. Report of a Working Group of the Royal Society into Risk. Royal Society: London. 1994

Presents the opposing scientific and social science viewpoints on risk. Dated. For students and specialists only.

CHAPTER 4

75. Barry Turner. *Man-Made Disasters.* Wykeham Publications. 1978
Brian Toft. *Learning From Disaster.* Leicester: Perpetuity Press. 2005
Andrew Weir. *The Tombstone Imperative.* London: Simon and Schuster UK. 2000

Turner's *Man-Made Disasters* is a landmark book and a must for students of risk. Brian Toft whom we met above (Chapter 3) worked with Turner at Exeter University. Toft's book is a wide-ranging examination of accidents from the management viewpoint. It is a must for managers, otherwise after an accident unpleasant questions about their diligence may be asked by the court. Andrew Weir is a writer and broadcaster. His book reflects his very interesting and productive research. It is warmly recommended to those interested in aviation, but not to nervous passengers.

78–9. James Reason. *Managing the Risks of Organisational Accidents.* Aldershot: Ashgate. 2005, p 71 and pp 140–1

James Reason's speciality is the examination of the situation in which people find themselves, combined with an analysis of the kind of mistakes which they may make. He has popularised the term 'latent error' and the Swiss cheese model of system failure. Visualise a row of slices of Swiss cheese of the holey variety; the event trajectory has to navigate all the way through the system via the holes to produce a system failure and an accident. Reason's books are interesting. His best known is *Human Error.*

81. Brookes, Andrew. *Disaster in the Air.* Shepperton: Ian Allan/Penguin. 1992, p 135

Andrew Brookes is an RAF pilot. He has written several well-informed and detailed books on aviation history and aviation safety.

85. Rasmussen, J. Human Error and the Problem of Causality in the Analysis of Accidents. *Phil. Trans. R. Soc.* Lond. B 327, 449–462 (1990)

Rasmussen is a thoughtful and well-regarded analyst of accident aetiology. Recommended to students of the subject.

85. Lewontin, R.C. The Doctrine of DNA. London: Penguin. 1993

86. Bob McKercher. *Phantom Demand: How Some Research 'Proves' Demand When None Really Exists.*

Bob McKercher is an Associate Professor at Hong Kong Polytechnic University.

87. Alexander McCall Smith. *Espresso Tales.* London: Abacus. 2008

Alexander McCall Smith was formerly Professor of Medical Law at the University of Edinburgh. He is a prolific and entertaining writer, the creator of Mma Ramotswe, traditionally-built lady detective of Botswana.

87. Michael Faure, Robert Young, Paul Fenn. *Acts and Omissions: Causal Uncertainty and the Efficiency of Tort Law.* Paper prepared for the EALE Conference, Nancy, September 18–20, 2003. My thanks to my colleague Rob Young, at the Nottingham University Business School, for drawing my attention to this paper, some of the ideas of which informed my research into causation.

CHAPTER 5

Many accounts of major disasters have been published; their contents obviously overlap. It is therefore not easy to be sure that one has given proper acknowledgement to all the sources from which one has drawn material. My compromise is to cite the works which are sitting on my desk; unfortunately I cannot do the same for the facts sitting in my memory, their origin forgotten.

95. Risk Transfer

Jerome Lederer: Aerospace Safety Pioneer and Flight Safety Foundation Organizer; Dies at 101. Flight Safety Foundation News 04–01.

The Flight Safety Foundation is an American organisation. A series of very able administrators have established it as the world's leading aviation safety body.

96. Ford Pinto

www.autosafety.org/article php?did=522&scid=8

The US Center for Auto Safety was founded in 1970 by Ralph Nader and the Consumers Union. It is a Washington DC based lobbying organisation, a valuable counterbalance to the exploitive nature of many of the major motor manufacturers, whose behaviour illustrates Malik's first two Laws of Life (Appendix to Chapter 15).

99. Titanic

Robert E Mittelstadt. *Will Your Next Mistake Be Fatal?* New Jersey: Wharton School Publishing. 2005

The Titanic disaster is discussed in many books. Professor Robert Mittelstadt's book is written for businessmen wanting to avoid 'the chain of mistakes which can destroy your organisation'. His accounts of some major disasters – Titanic, Challenger, Enron and others – include details which are useful and relevant to our analysis. He is the principle reference for our discussion of the *Titanic.*

101. R101

Stanley Stewart. *Air Disasters.* Shepperton: Ian Allan: Penguin. 1986
James R Chiles. *Inviting Disaster.* New York: Harper Business. 2002

These two books are the principle references for our discussion of the R101.

Air Disasters gives valuable detail on some dozen major air accidents. Stanley Stewart himself gives particular credit to Sir Peter Masefield's *To Ride the Storm: the Story of the R101* and to Sir Peter's article on the R101 accident in the *Aeronautical Journal.*

Professor James Chiles' *Inviting Disaster* is thoughtful. It is organised usefully into chapters each addressing an accident source, with a detailed examination of major accidents emanating from that source.

104. The Trident

Stanley Stewart. *Air Disasters.* Shepperton: Ian Allan/Penguin. 1986

107. *Challenger*

James R Chiles. *Inviting Disaster.* New York: Harper Business. 2002

Principle reference and source of the descriptive words 'ultimate cutting torch'.

109. Korean 747

Air Accident Investigation Branch. *Aircraft Accident Report 3/2003*

111. Tenerife 747

Stanley Stewart. *Air Disasters.* Shepperton: Ian Allan/Penguin. 1986
Karl E. Weick. *The Vulnerable System: An Analysis of the Tenerife Air Disaster.* pp 173–198 in New Challenges to Understanding Organizations, editor K Roberts New York: Macmillan

113. DC–10 Slats

Stanley Stewart. *Air Disasters.* Shepperton: Ian Allan/Penguin. 1986
Charles Perrow. *Normal Accidents.* Princeton University Press. 1999

CHAPTER 6

119. Ocean Ranger

Chiles, James R. *Inviting Disaster.* pp 17–36

119. Bhopal

Chiles, James R. *Inviting Disaster.* pp 259–68

120. USS *Vincennes*

Gero, David. *Flights of Terror.* Yeovil: Patrick Stephens. 1997, p 114

121. Piper Alpha

Reason, James. Managing the Risks of Organisational Accidents, p 89

123. *Herald of Free Enterprise*

Report of Court No 8074, *mv Herald of Free Enterprise.* DOT, 1987

125. Valujet

Langewiesche, William. *The Lessons of ValuJet 592.* Atlantic Monthly. 1998 Mar 15 pp 81–98

126. Air France Concord

Weir, Andrew. *The Tombstone Imperative,* p 310 and p 320

127. NHS

Toft, Brian. *External Inquiry into the adverse incident that occurred at Queen's Medical Centre Nottingham 4th January 2001*

CHAPTER 7

141. *Scott Snook. Friendly Fire.* Princeton University Press. 2000

149. *Friendly Fire* p 7. Pentagon quoted in *Aerospace Daily,* 1994:81

152. *Friendly Fire* p 108

152. Friendly Fire pp 202–203

154. RAF Chinook ZD 576

RAF Proceedings of a Board of Inquiry into the Accident to Chinook HC2 ZD 576 on 2 June 1994. All grammar *sic.*

157. AP 3207 8th Edition, Annex E to chap 8, para 9. 8-G–3

157. Report of the Committee on Civil Aircraft Accident Investigation and Licence Control. CAP169. HMSO, 1961

158. Statement to Select Committee of the House of Lords on Chinook ZD 576

159. WH Tench. Report of Study of Aircraft Accident Investigation Procedures in the Armed Services. Jan 1987

159. Private letter from Bill Tench to Captain John Cook, 13 January 2000

161. Report of the Select Committee of the House of Lords on Chinook ZD 576, para 174

161. The Government's Response to the Report from the Select Committee on Chinook ZD 576, 1.5, 22 July 2002

161. Lords Hansard 30 Apr 2001: Column 1589

CHAPTER 8

176. NHS Accident

Toft, Brian. *External Inquiry into the adverse incident that occurred at Queen's Medical Centre Nottingham 4th January 2001*

Blackhawk Shoot Down

181. *Friendly Fire*, p 136–38

181. *Friendly Fire*, p 214–15

RAF Chinook ZD 576

184. *RAF Board of Inquiry into Chinook ZD 576*, p 4–3

In re the accident to RAF Chinook ZD 576, further good reading is the psychologist, Professor Norman Dixon. In his book *On the Psychology of Military Incompetence,* he examines the blocks on open-mindedness and learning to often characteristic of an authoritarian personality.

184. 'Helicopters 500 times more dangerous … ' Report of a Royal Society Study group. *Risk: Analysis, Perception and Management.* London: The Royal Society, 1992.

CHAPTER 9

205. Adams, John. *Risk.* London: Routledge. 2001. p. 50

205. Summers storm: *a feminist show trial at Harvard.* Christina Hoff Summers. National Review, Feb 14 2005

206. Wall Street Journal. *The professor who's suing our students.* Joseph Rago. Extract in THE WEEK, 17 May 2008, p 13

CHAPTER 10

212. Tim Harford. *The Logic of Life.* London: Little Brown. 2008

To most of us, rational decision-making is a product of common-sense. Tim Harford appears to claim it as a product of economics. Therefore Boo to economists. *The Logic of Life* is nonetheless an interesting and very readable contribution to our understanding of decision-making. Another economist, Professor Paul Ormerod, in his book *Why Most Things Fail,* argues that failure is the natural default mode of life, that there is an Iron Law of Failure. You may find his book a useful counterpoint to our own study, of how (to try) to ensure that things do not fail.

CHAPTER 11

234. Colonel John Hughes-Wilson. *Military Intelligence Blunders.* London: Robinson. 1999

Excellent book, apart from its dismissive view of cognitive dissonance.

234. Max Bazerman. *Why Good Accountants Do Bad Audits.* Harvard Business Review. November 2002.

235. Richard H Thaler and Cass R Sunstein. *Nudge.* Yale. 2008

236. Frank Tallis. *Hidden Minds.* London: Profile Books. 2002

236. Gerd Gigerenzer. *Gut Feelings.* London: Penguin. 2007

If you wish to go deeper into sensemaking and decisions, have a look at Professor Karl Weick's *Sensemaking in Organizations.* It is an academic rather than an easy book, but full of interesting observations.

CHAPTER 12

244. The section on Pathways draws extensively on Gigerenzer's *Gut Feelings.* Pp 169–77

245. Gigerenzer's Figure 9-2, Fast and Frugal Tree, p 174, draws on Green L.A., and D.R Meyr, *What alters physicians' decisions to admit to the coronary care unit?* The Journal of Family Practice 45:210–26

249. *Gut Feelings.* p. 3

CHAPTER 13

258. Scott Sagan. *The Limits of Safety.* Princeton University Press. 1993, p 99

Scott Sagan's book is subtitled *Organizations, Accidents and Nuclear Weapons.* It is a detailed, fascinating and very sobering account of the US handling and mishandling of its thermonuclear armoury. An everyday story of muddle and finger-trouble, but with a difference. The finger is on the button to oblivion. One can conclude only that the longer that such weapons exist and the more incompetent, unstable or corrupt the countries that have them, the more inevitable is a nuclear accident or event.

258. Sunday Times. *US Hits Panic Button.* Oct 21, 2007. p1.24

260. James Montier. *Seven Sins of Fund Management.* www.drkresearch.com. 2005. p1

265. Robert G. Hagstrom. *Investing, The Last Liberal Art.* London: Texere. 2000

267. Nick Maddocks. *The West at War 1939–45.* Stroud: Sutton. 2005

Stephen Bungay. *The Most Dangerous Enemy.* London: Aurum Press. 2001

The West at War 1939–45 contains personal, very frank accounts of war service. Particularly interesting are soldiers' accounts of the British army's unpreparedness for war, and the blunt views of my great friend, the late Keith Aldridge, on the RAF's unpreparedness. It sent many pilots into the Battle of Britain with little or no experience of their aircraft, and with absolutely zero combat training. Keith was sent into combat with less than three hours on the Hurricane; he was shot down and badly burned. In his masterly account of the Battle of Britain, Stephen Bungay likened the RAF's use of untrained pilots to entering a learner driver for a Grand Prix, but with the small difference that the other drivers will be trying to kill him.

267. Hilaire Belloc. *Travellers Tales,* in Lord Wavell, *Other Men's Flowers.* London, Jonathan Cape, 1854, p 77.

267. Colonel Alfred Daniel Wintle MC. *The Last Englishman.* Michael Joseph: London. 1968.

This book, now difficult to obtain, describes the splendid Colonel Wintle's lifetime of war with clots, crooks, and bureaucrats. He is particularly pungent in his comments on jobsworth staff and intelligence officers in the 1930's. There is a memorable short television film of his exploits, *The Last Englishman,* with Jim Broadbent as Colonel Wintle.

CHAPTER 14

273. Lord Owen. *Inside Blair's Brain. Sunday Times,* March 16 2008, p 4.1

4. Simon Jenkins. *The rise of imperialism lite is prolonging the Iraqi horror. Sunday Times,* March 16 2008, p 1.16

282. Lord Wavell. *Other Men's Flowers.* London: Jonathan Cape. 1954. p 414

288. Drew Westen. *The Political Brain.* New York: Public Affairs. 2007. p 145, citing

Jost. J.T, Glaser, J., Kruglanski, A.W., & Sulloway, F. J. (2003). Political Conservatism as Motivated Social Cognition. *Psychological Bulletin, 129, 339–375*

293. Richard Brooks. *Biteback.* Sunday Times, March 16 2008, Culture Section, p 14

CHAPTER 15

305. Border force to shut at weekends. Sunday Times, March 2 2008. p 1.5

306. Karl E Weick. *Sense-Making in Organisations.* London: Sage. 1995, p 137

313. Lauren Slater. *Opening Skinner's Box: Great Psychological Experiments of the 20th Century.* London: Bloomsbury. 2004. Extract: The Man Who Made Fools of the Psychiatrists. The Week, 7 February 2004, p 40–1.